2nd International Conference on Upstream Engineering and Flow Assurance 2014

Topical Conference at the 2014 AIChE Spring Meeting and 10th Global Congress on Process Safety

New Orleans, Louisiana, USA
30 March - 3 April 2014

ISBN: 978-1-63439-066-8

Printed from e-media with permission by:

Curran Associates, Inc.
57 Morehouse Lane
Red Hook, NY 12571

Some format issues inherent in the e-media version may also appear in this print version.

Printed by Curran Associates, Inc. (2014)

AIChE
3 Park Avenue
New York, NY 10016-5991

Phone: (203) 702-7660
Fax: (203) 775-5177

www.aiche.org

Additional copies of this publication are available from:

Curran Associates, Inc.
57 Morehouse Lane
Red Hook, NY 12571 USA
Phone: 845-758-0400
Fax: 845-758-2634
Email: curran@proceedings.com
Web: www.proceedings.com

TABLE OF CONTENTS

354012 Possible vs. Practical: Engineers Must Lead the Development of Practical Technologies

Abstract:

Chemical Engineering is need today more than ever. There is a strong societal desire for sources of energy and feedstocks that can sustainably support a high standard of living for all. Thermodynamic, material and economic challenges stand in the way. Most of the general public does not have the basic understanding of the energy flows, chemical transformations, and the scale of the systems that support our current way of life. These concepts are fundamental to chemical engineering. Chemical engineers should be on the vanguard of separating hype from truly practical options yet occationally we let society down. This talk will present several examples where the fundamentals of chemical engineering can be used to access the practicality of proposed solutions to energy and water issues, as well as explain why the venture capital model struggles in the materials market.

352091 A Model for Three Phase Flow in Horizontal Pipes

Abstract:

Experimental observations on three phase flow of gas and two immiscible liquids in pipes is very limited inspite of its prevalence in the oil and gas industry. A few investigators reported observations on the regime maps, phase hold-ups and pressure drop. Extension of the two phase flow models to explain three phase flows by defining two out of the three phases as a pseudo-homogeneous phase was inadequate. In this paper, an equation for the pressure drop in three phase flow through horizontal pipes is derived by extending the separated flow model for the three phases in terms of Lockhart-Martinelli parameters. The model is validated with the experimental data available in the literature.

353112 Immiscible Displacement in Homogeneous and Heterogeneous Porous Media: Quantifying Interfacial Phenomena

Abstract:

Immiscible displacement is ubiquitous in many subsurface multiphase flow systems and pore-scale characterization of these processes aids in the development of quantitative, empirical relationships required in Darcy-scale models. Here, we present experiments in which a wetting fluid is displaced by a non-wetting fluid in homogeneous and heterogeneous porous micromodels. The resulting steady-state fluid distributions are analyzed to characterize fluid-fluid and fluid-solid interfacial length in terms of saturation, fingering pattern, capillary number, and porosity. A comparison between the homogeneous and heterogeneous experiments demonstrates that heterogeneity expands the range of conditions under which capillary fingering is observed. In all micromodels, the number of non-wetting isolated fingers (i.e., blobs and ganglia) increased within the range of injection flow rates. The results show a clear relationship between interfacial length and saturation; however we did not observe a clear relationship between interfacial length and fingering pattern. In addition, once saturation was taken into account, interfacial length was independent of the capillary number. A clear relationship is observed between the non-wetting–wetting fluid interfacial length, non-wetting fluid saturation, and porosity. We anticipate our results will serve as model to quantify relationships for interfacial phenomena in pore-scale drainage and help with the development of Darcy-scale models.

347844 Comparison Between Two-Phase Flow Models and Wire Mesh Sensor Measurements in Medium and Large Diameter Pipes

Abstract:

Wire mesh sensors (WMS) are recently developed fast imaging instruments that have been used to investigate two-phase flows. WMS measure either the electrical conductivity or capacitance of the two phases, allowing the local void fraction to be inferred [1, 2]. The WMS have been benchmarked against the gamma densitometer, a high-resolution instrument widely used in the industry for two-phase flow metering; therefore providing confidence with experimental data [3]. In this work, we compare the results of an extensive WMS experimental campaign,with a number of two-phase flow models based on the drift-flux approach [4]. Void fraction is a critical parameter in two-phase flows. Drift-flux models are semi-mechanistic models that compare to fully mechanistic models use an explicit equation for the void fraction, thus reducing computational time significantly. Moreover, most mechanistic models cannot capture all the flow patterns and pipe inclinations within a single framework without adding complexity and computational overhead. A number of authors have developed drift-flux models for a range of flow patterns and inclinations which seem suitable for bubbly, slug and churn flows [5, 6].

The experiments consisted of WMS measurements using water/air and oil/air in medium and large diameter pipes with pipe inclinations ranging from vertical to horizontal. We compare the performance of several drift-flux closure equations developed for a range of flow patterns and inclinations. The models have been ranked and the empirical parameters of the best performers have been optimized. Additional comparisons between the final optimized models, the experimental data and the mechanistic model OLGA-S are also presented in this study.

[1] Prasser et al. (1998) A new electrode-mesh tomograph for gas-liquid flows, *Flow Meas. Instrum.,*9, 111-119

[2] Da Silva et al. (2007) Capacitance wire-mesh sensor for fast measurement of phase fraction distribution, *Meas. Sci. Technol.*, 18, 2245-2251

[3] Sharaf et al. (2011) Comparison between wire mesh sensor and gamma densitometry void measurements in two-phase flow, *Meas. Sci. Technol.*, 22, 1-13

[4] Ishii and Hibiki (2006) Thermo-fluid Dynamics of Two-Phase Flow, Springer

[5] Shi et al. (2005) Drift-flux Modeling of Two-Phase Flow in Wellbores, *SPE J.*, 24-33

[6] Choi et al. (2012) An Efficient Drift-flux Closure Relationship to Estimate Liquid Holdups of Gas-Liquid Two-Phase Flow in Pipes, *Energies*, 5, 5294-5306

351674 Experimental Flow Loop Setup Used to Investigate Hydrate Formation in Water-in-Oil Emulsions

Abstract:

Gas hydrates are non-stoichiometric crystalline solids with a characteristic guest – host structure that develop from mixtures of water and low molecular weight hydrocarbons. The agglomeration and plug formation tendencies of hydrates in oil and gas flow lines and the ensuing capital losses have provided an impetus to develop reliable, efficient, and economical flow assurance strategies. In this work, we present an experimental setup that is used to quantify emulsion behavior and hydrate formation in flowing systems. The flow loop (8 ft X 8 ft) consists of jacketed 1 in. tubing. Fluid is circulated in the flow loop using a 5 hp progressive cavity pump. The process fluid temperature is controlled in the range (-30°C to 150°C). Flow measurement is accomplished using a mass flow meter, and an inline microscope provides quantitative drop size distributions of emulsions as well as hydrate particle size. The microscope also provides real time images of hydrate slurry flow, adhesion of hydrate particles to inner pipe walls, and eventual plug formation. Quantitative droplet/particle size data are utilized to establish relationships between emulsion droplet size and hydrate behavior in flowing conditions for flow assurance applications.

astro technology

ADVANCED DEEPWATER MONITORING

Reza Asgharzadeh Shishavan

John Hedengren

LAND | DEEPWATER | AEROSPACE | astrotechnology.com

astro technology

ABOUT ASTRO TECHNOLOGY

ADVANCED INSTRUMENTATION FOR:

- Subsea fields
- Pipelines and risers
- Space structures
- Rocket Motors

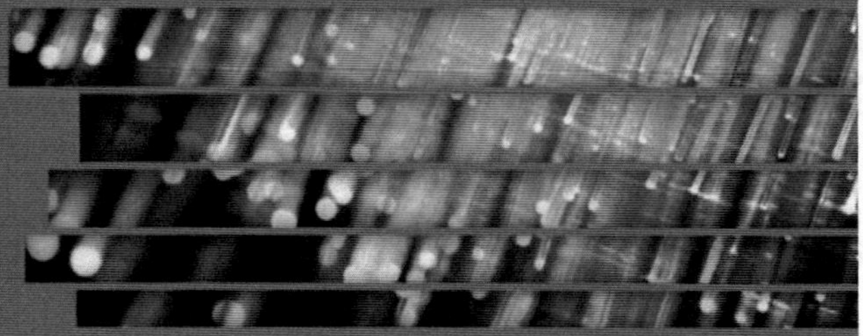

ENGINEERING CAPABILITIES INCLUDE:

- System integration
- Real-time embedded systems
- Experimental stress analysis
- Fiber-optic sensor technology
- Conventional sensor integration
- Environmentally hardened systems
- Software development

LAND | DEEPWATER | AEROSPACE | astrotechnology.com

FAILURES TO MONITOR AND PREDICT

- Detect early warning signs
- Automate monitoring of critical systems
- Give critical data to key decision makers

Deepwater Horizon
2010

Thunder Horse
2005

Texas City Refinery
2005

DEVIL'S TOWER BASS LITE

- Located in Atwater Valley Block 426
- Utilizes 20.3-cm (8-in) diameter flowline
- 90-km (56-mi) length
- Ties to Devil's Tower in Mississippi Canyon
- Production – Up to 130 million cubic feet per day
- Water depth – 2,050 m (6,750 feet)
- Commenced operation in February 2008

MONITORING:

- Pressure
- Temperature
- Hoop and Axial Strain

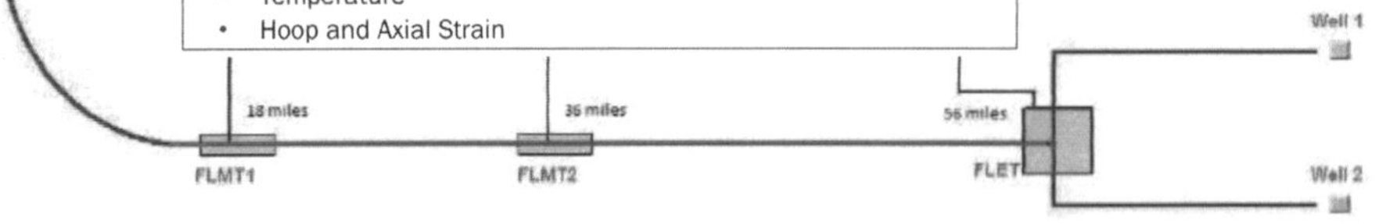

LAND | DEEPWATER | AEROSPACE | **astrotechnology.com**

astro technology

PREVIOUS INSTRUMENTATION ON DEEPWATER RISERS

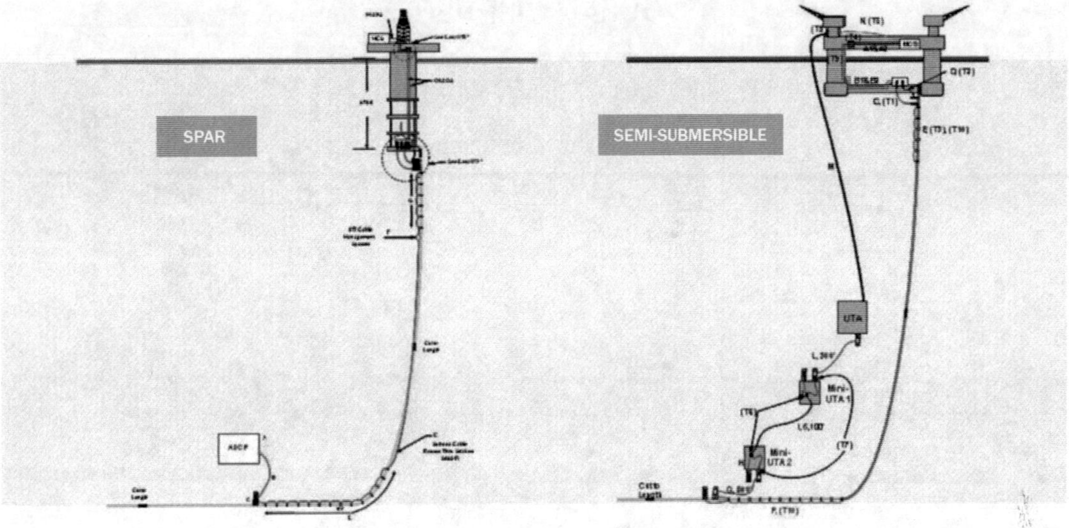

LAND | DEEPWATER | AEROSPACE | **astrotechnology.com**

astro technology

TROIKA — GULF OF MEXICO

ASTRO TECHNOLOGY pioneered the use of fiber-optic sensors on a subsea pipeline to monitor pressure, strain and vibration in external casing pipe bundle during fabrication.

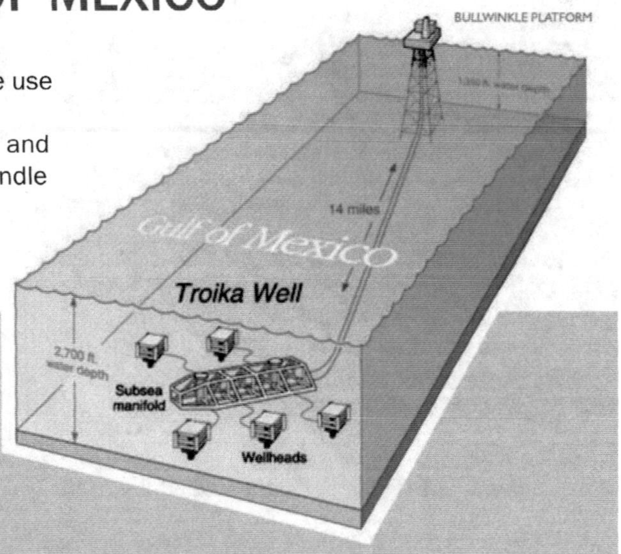

LAND | DEEPWATER | AEROSPACE | **astrotechnology.com**

astro technology

FIBER-OPTIC SENSORS FOR DEEPWATER DRILLING
OCEAN CLIPPER

LAND | DEEPWATER | AEROSPACE | **astrotechnology.com**

astro technology

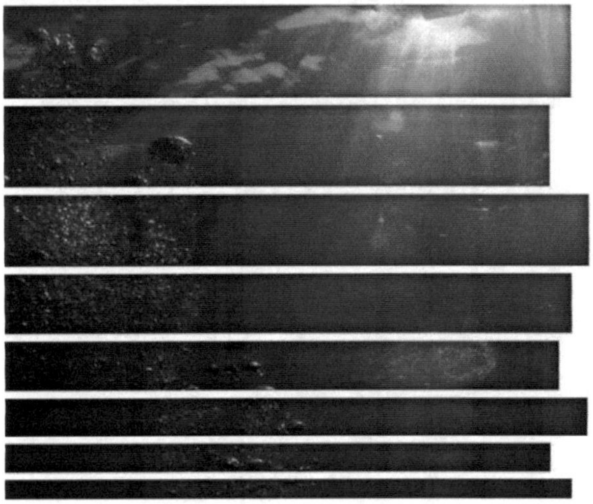

OVERVIEW *of* INSTRUMENTATION METHODS

TRIDENT
SUBSEA SYSTEMS
ASTRO TECHNOLOGY
DEEPER INSIGHT.

LAND | DEEPWATER | AEROSPACE | **astrotechnology.com**

astro technology

FIBER BRAGG GRATINGS

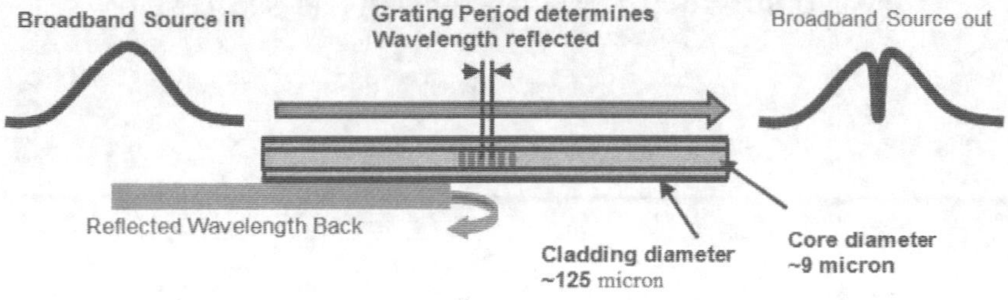

Relationship between Strain (ε) and Change in Wavelength ($\Delta\lambda_b$)

$\Delta\lambda_b / \lambda_b = (1-p_e)\varepsilon$, where p_e is the photoelastic constant for glass and λ_b is the base wavelength

Multiple Gratings (sensors) can be placed on a single fiber, enabling high sensor count per fiber channel.

TRIDENT SUBSEA SYSTEMS

LAND | DEEPWATER | AEROSPACE | astrotechnology.com

astro technology

RELATIONSHIP BETWEEN WAVELENGTH & STRAIN

EXAMPLE BASE WAVELENGTH FOR A SINGLE FBG

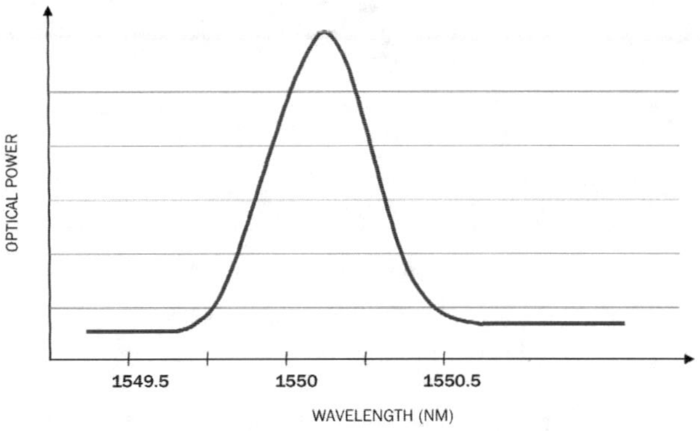

TRIDENT SUBSEA SYSTEMS

LAND | DEEPWATER | AEROSPACE | astrotechnology.com

astro technology

RELATIONSHIP BETWEEN WAVELENGTH & STRAIN — TENSION

REFLECTED SIGNAL FROM THE PREVIOUS FBG IN TENSION

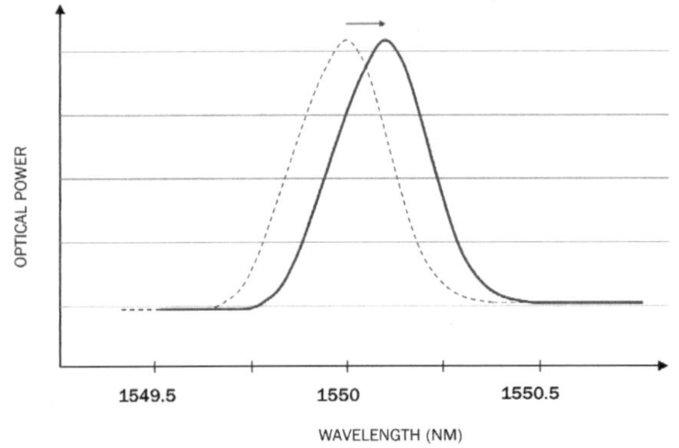

TRIDENT

LAND | DEEPWATER | AEROSPACE | **astrotechnology.com**

astro technology

RELATIONSHIP BETWEEN WAVELENGTH & STRAIN — COMPRESSION

REFLECTED SIGNAL FROM THE PREVIOUS FBG IN COMPRESSION

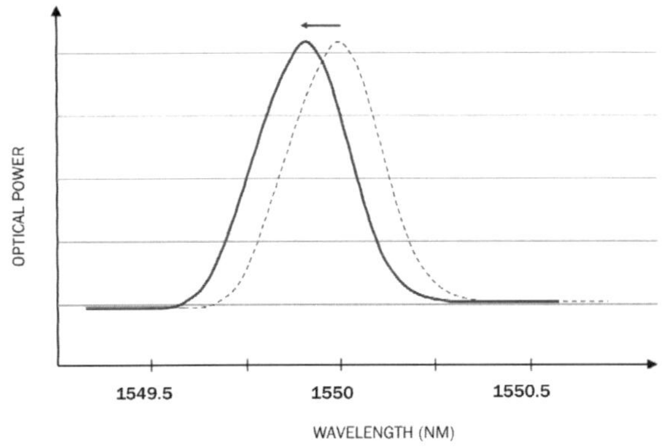

TRIDENT

LAND | DEEPWATER | AEROSPACE | **astrotechnology.com**

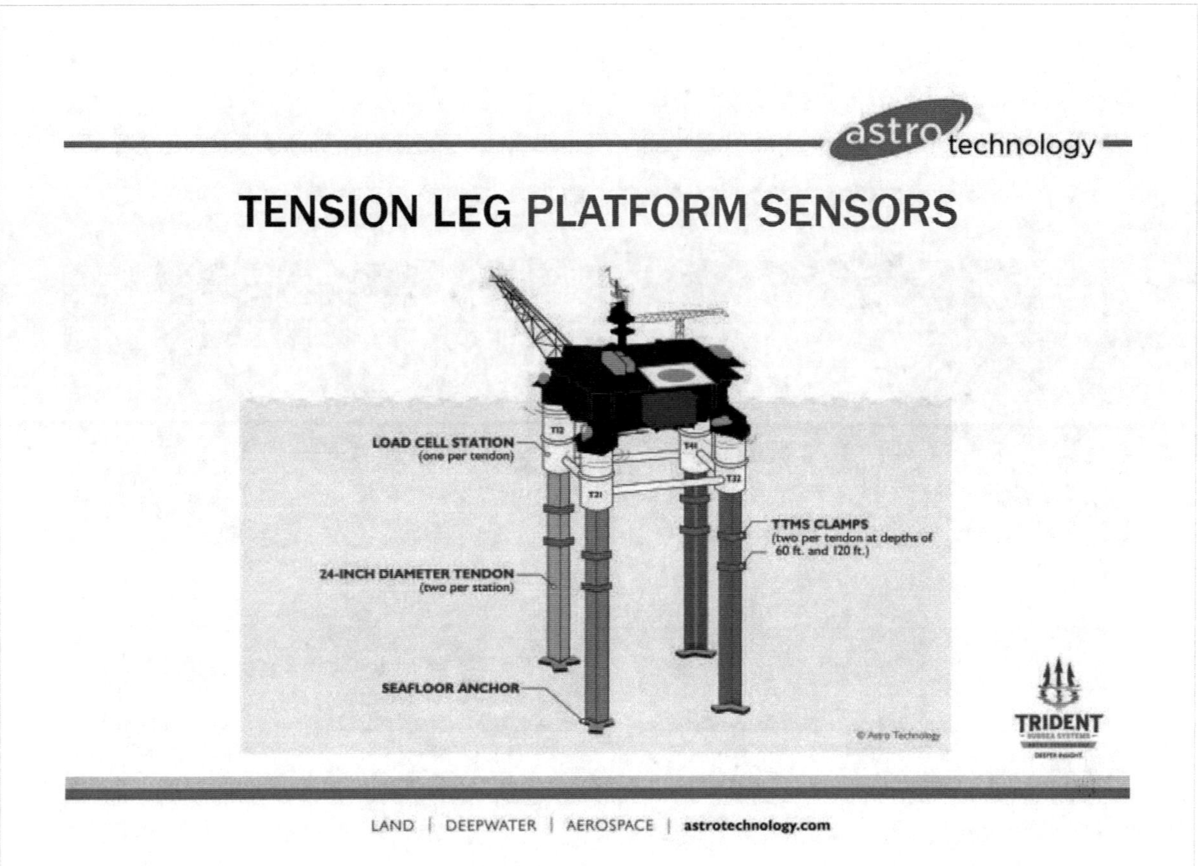
astro technology
TENSION LEG PLATFORM SENSORS
LOAD CELL STATION
(one per tendon)
TTMS CLAMPS
(two per tendon at depths of 60 ft. and 120 ft.)
24-INCH DIAMETER TENDON
(two per station)
SEAFLOOR ANCHOR
© Astro Technology
TRIDENT
LAND | DEEPWATER | AEROSPACE | astrotechnology.com

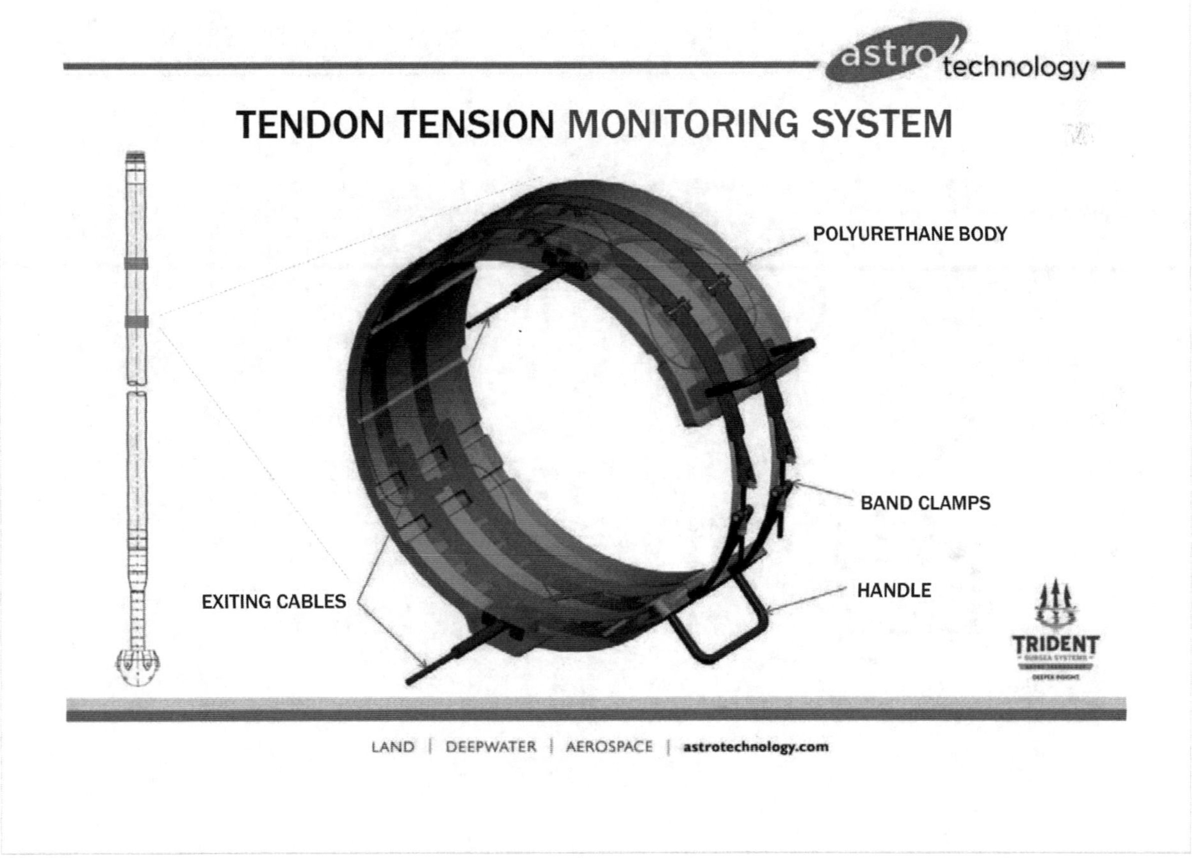
astro technology
TENDON TENSION MONITORING SYSTEM
POLYURETHANE BODY
BAND CLAMPS
HANDLE
EXITING CABLES
TRIDENT
LAND | DEEPWATER | AEROSPACE | astrotechnology.com

astro technology

TENDON BAND PREPARATION

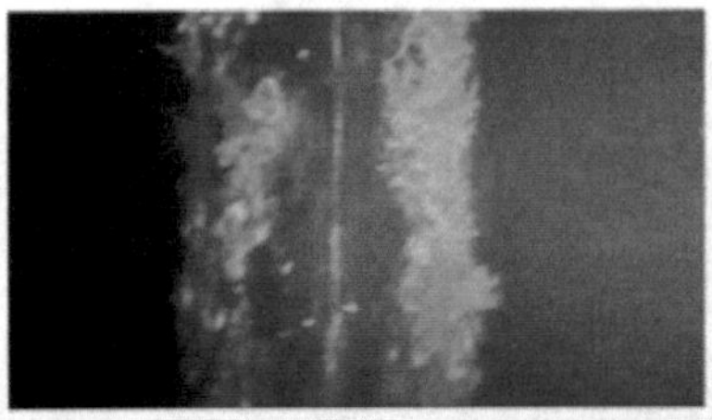

Marine Growth
(BEFORE)

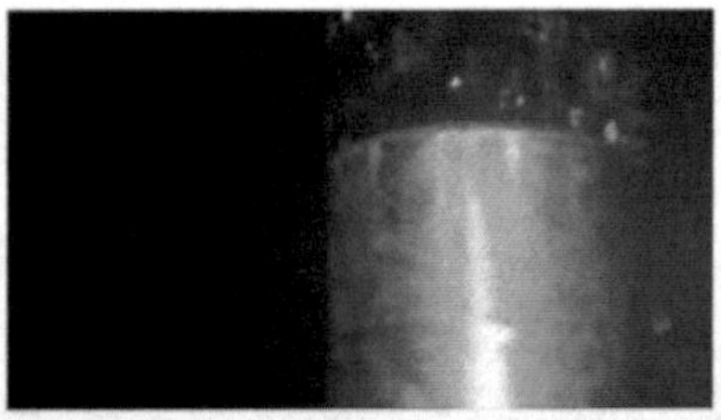

Clean Band
(AFTER)

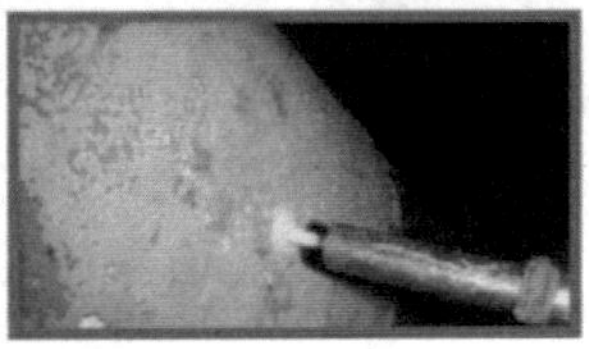

Cleaning with Water Jet

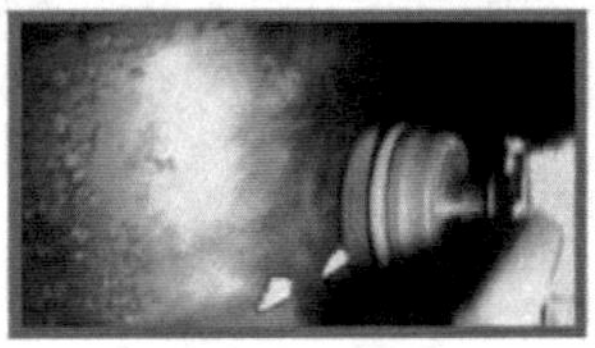

Polishing to Bare Metal

TRIDENT

LAND | DEEPWATER | AEROSPACE | **astrotechnology.com**

astro technology

DIVER INSTALLATION

Diver with Clamp

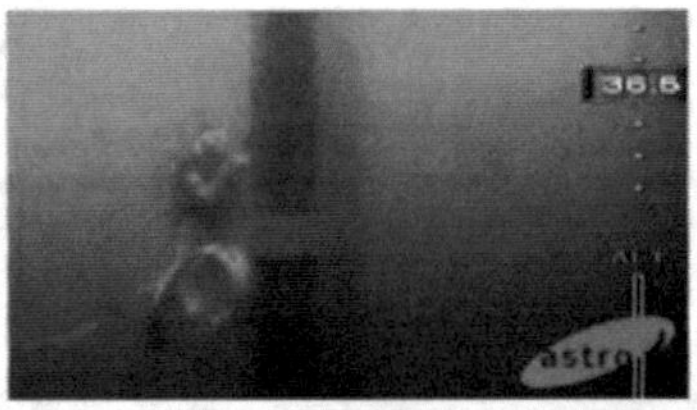

Riser Preparation

Clamp Installation

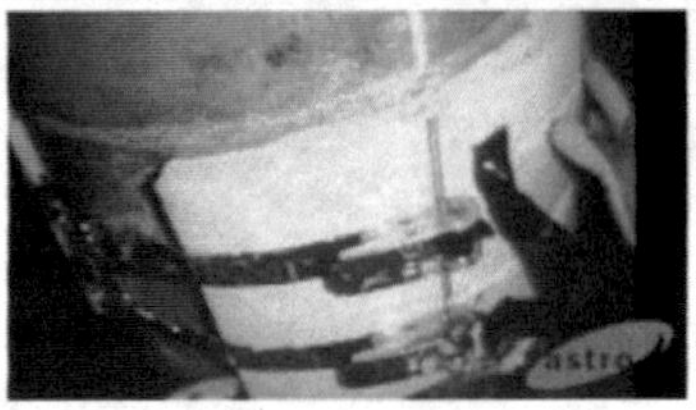

Clamp Inspection

TRIDENT

LAND | DEEPWATER | AEROSPACE | **astrotechnology.com**

TEMPERATURE & STRAIN GAUGES

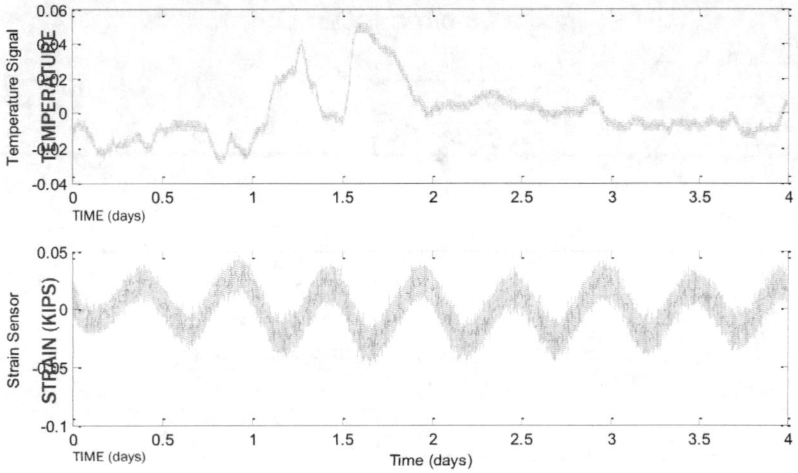

LAND | DEEPWATER | AEROSPACE | **astrotechnology.com**

COMPENSATE FOR DYNAMIC TEMPERATURE VARIATION

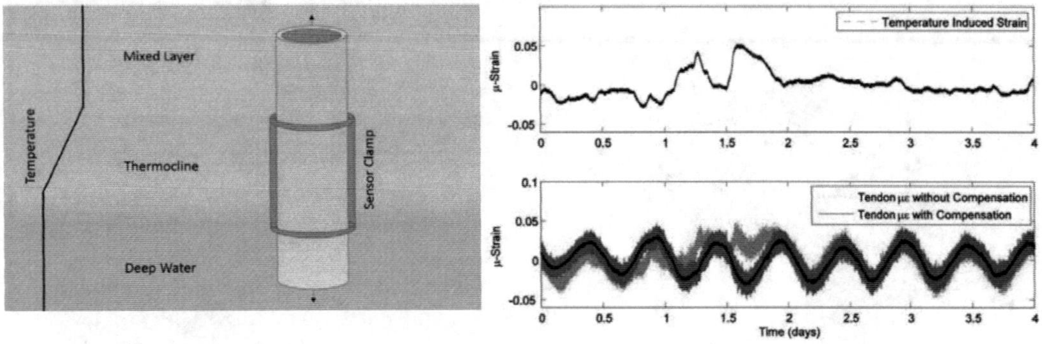

LAND | DEEPWATER | AEROSPACE | **astrotechnology.com**

TRIDENT MONITORING SYSTEM

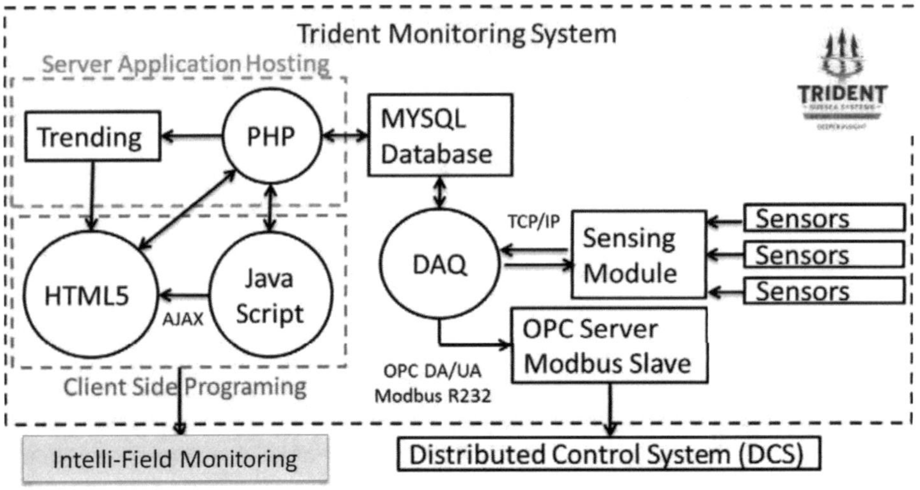

ENSURE SUBSEA BONDING STRENTH

Button Pull Tests

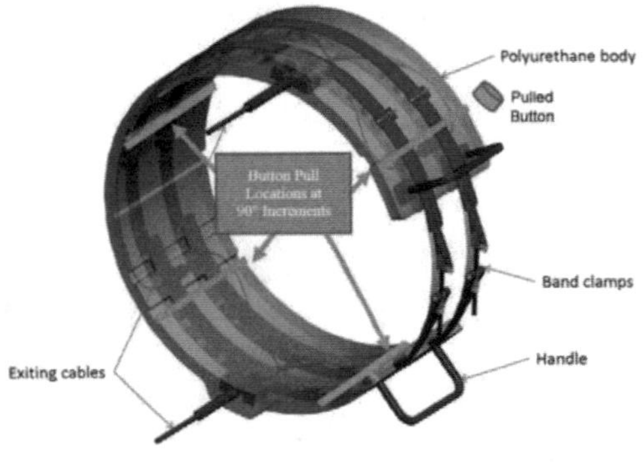

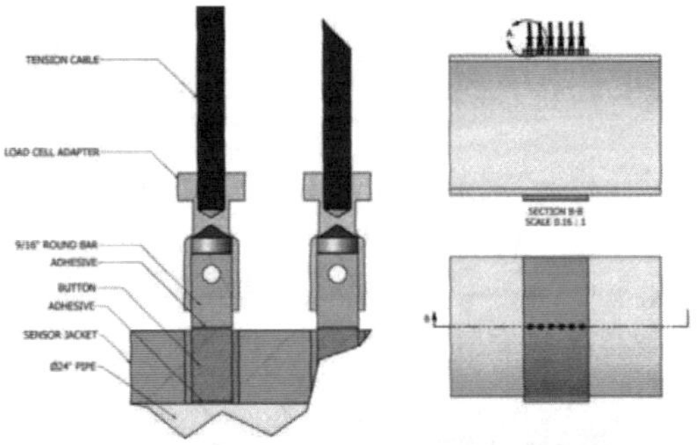

astro technology

BONDING MAINTAINED IN ACCELERATED AGING

Clamp	Avg (psi)	StDev (psi)	Min (psi)	Max (psi)	Description
BB	**292.0**	108.9	113.6	498.6	**Dry Bonded Control**
AA	**81.4**	47.2	31.1	193.5	**Wet Bonded Control**
A	94.0	30.4	45.0	155.8	Four Point Bending (Left)
B	174.1	112.4	29.8	503.3	Four Point Bending(Left)
E	**91.1**	46.9	49.5	150.8	**Four Point Bending (Center)**
C	142.5	65.6	45.8	267.9	Four Point Bending (Right)
D	136.3	73.6	57.5	358.7	Four Point Bending (Right)
F	**114.0**	53.7	33.6	225.7	**Tension**
G	**105.3**	59.8	29.7	268.8	**Compression**

astro technology

IMPROVED CLAMP DESIGN

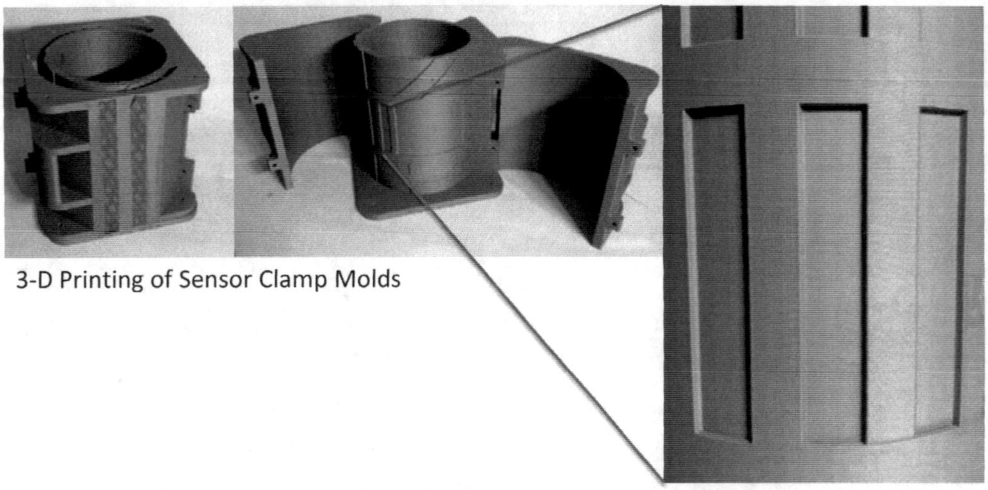

3-D Printing of Sensor Clamp Molds

LAND | DEEPWATER | AEROSPACE | astrotechnology.com

astrotechnology

BOAT DOCKING

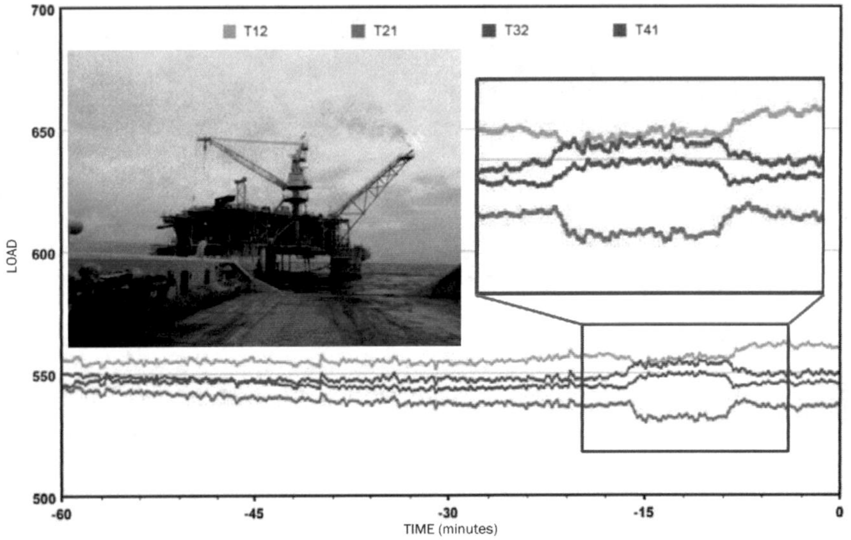

LAND | DEEPWATER | AEROSPACE | **astrotechnology.com**

astrotechnology

WEST AFRICAN ENVIRONMENT:

SUDDEN AND POTENTIALLY VIOLENT SQUALLS

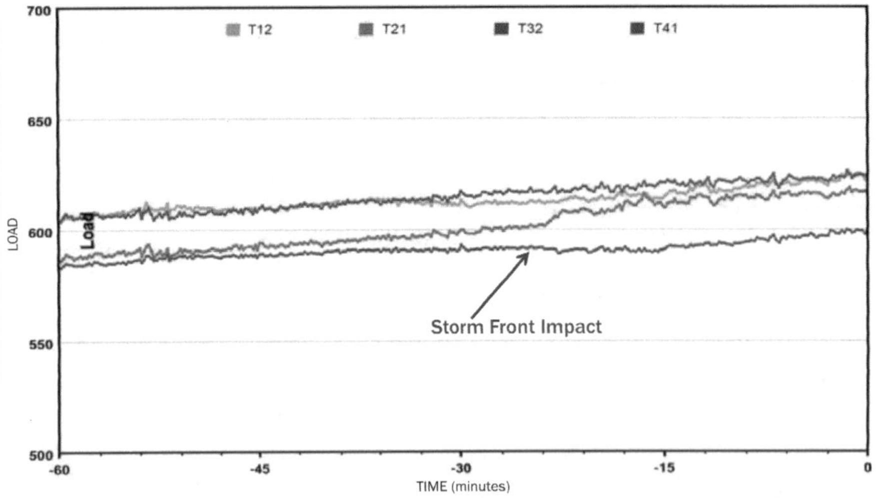

LAND | DEEPWATER | AEROSPACE | **astrotechnology.com**

SENSOR CALIBRATION WITH TIDE CYCLES

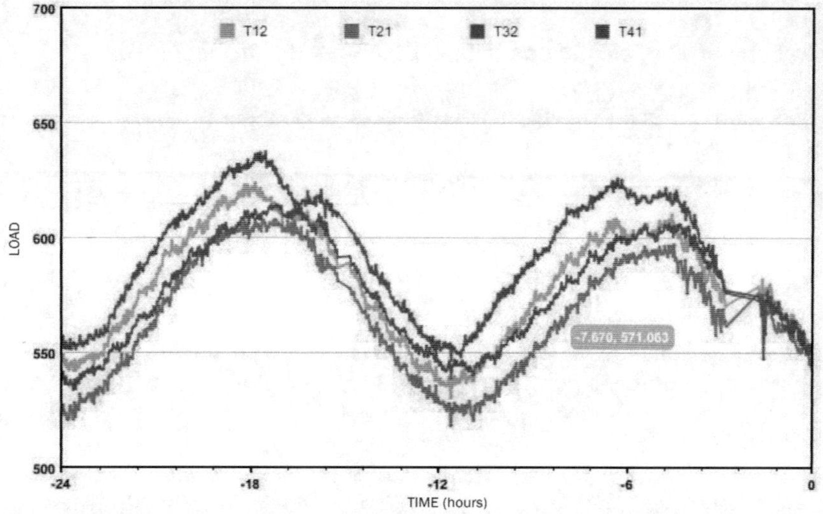

LAND | DEEPWATER | AEROSPACE | astrotechnology.com

ADVANCED DEEPWATER MONITORING SUMMARY

- Flow Lines (Temperature, Pressure, Hydrates)
- Risers, Tendons (Strain, Fatigue, Vibration)
- New Data Acquisition Overview
  - Calibration on Post-Installed Systems
  - Key Software Architecture Elements
- Observing Unique and Interesting Events
- Synthesizing Data into Actionable Information
  - Delivered to Key Decision Makers
- Future Activities Include Automation of Load Balancing and Abnormal Event Detection

LAND | DEEPWATER | AEROSPACE | astrotechnology.com

CLEAR GULF JOINT INDUSTRY PROJECT (JIP)

- Collaboration between oil and gas industry, NASA and Astro Technology

PROPOSED PARTNERS INCLUDE:

ExxonMobil, Apache, ConocoPhillips, Chevron, HESS, Shell, woodside, Anadarko Petroleum Corporation, Marathon Oil, PEMEX, PETROBRAS, Eni, bp, TOTAL, Statoil

- Create cutting-edge techniques for managing production
- Develop environmental and safety systems for drilling and production
- Respond to challenges faced when working in remote and harsh environments
- 9 specific tasks

LAND | DEEPWATER | AEROSPACE | **astrotechnology.com**

ADDITIONAL CONTENT

LAND | DEEPWATER | AEROSPACE | **astrotechnology.com**

technology

INSTRUMENTATION OF NASA'S ROBONAUT HAND

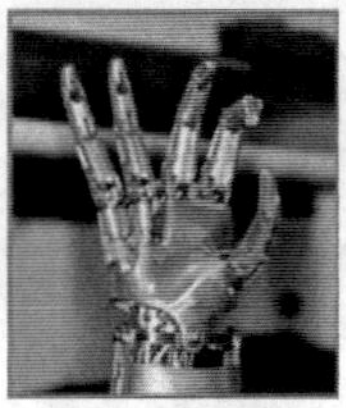

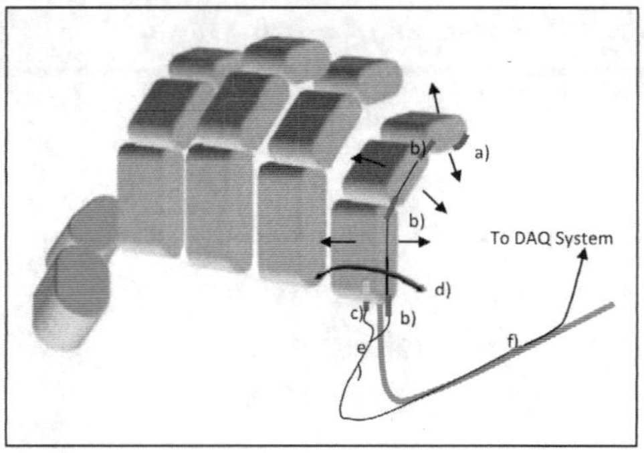

LAND | DEEPWATER | AEROSPACE | **astrotechnology.com**

technology

Space Shuttle Return-to-Flight

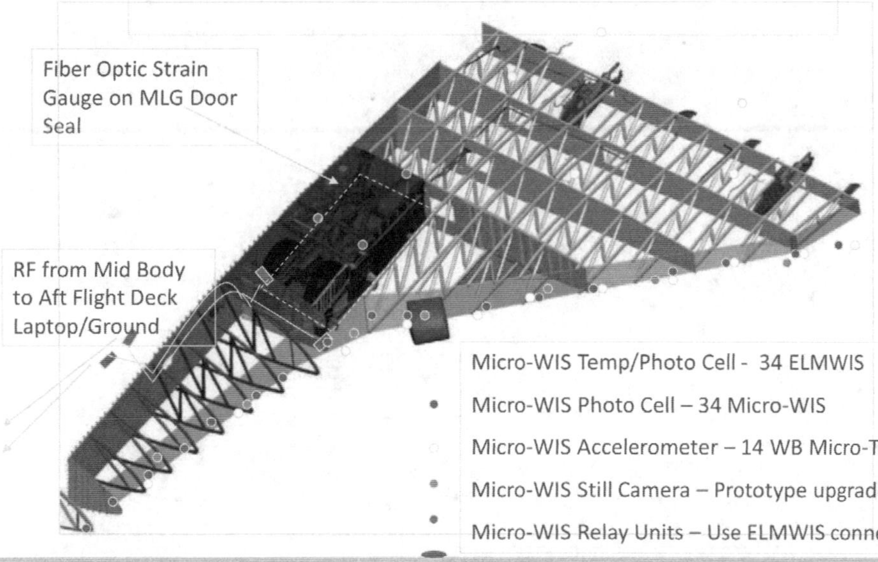

LAND | DEEPWATER | AEROSPACE | **astrotechnology.com**

astro technology

Solid Rocket Motor Test

- Apply fiber-optic sensor technology to solid rocket motors
  - In-situ sensors on new motors
  - Sensors bonded to the interior of existing motors
    - Fiber Bragg-gratings
    - Fabry Perot
- Measure mechanical and chemical properties during handling, storage, long-term aging, motor firing, case burst and damage assessment
- Provide tool for service life evaluation
- Early detection of possible structural failure

LAND | DEEPWATER | AEROSPACE | astrotechnology.com

astro technology

FIBER-OPTIC SENSOR ROCKET MOTOR TEST

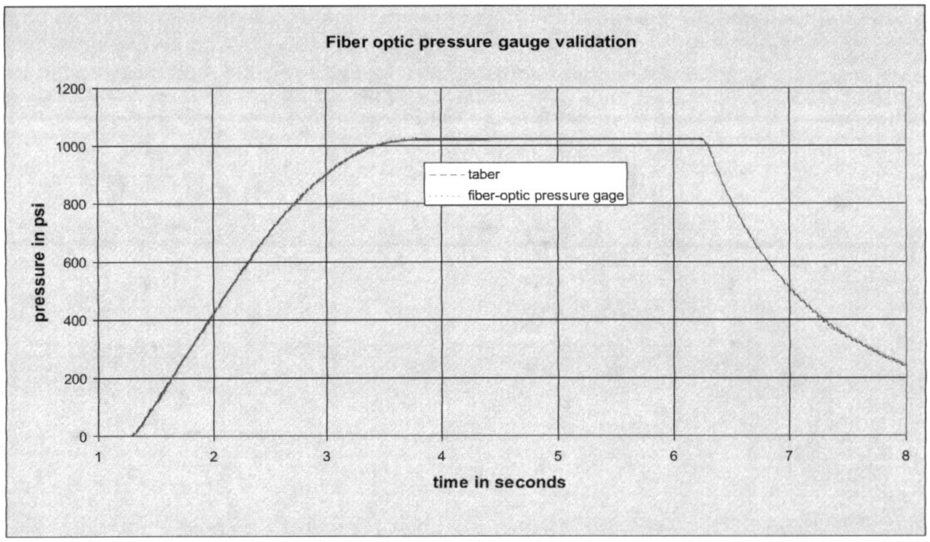

LAND | DEEPWATER | AEROSPACE | astrotechnology.com

astro technology

FIBER-OPTIC SENSORS FOR CIVIL STRUCTURES

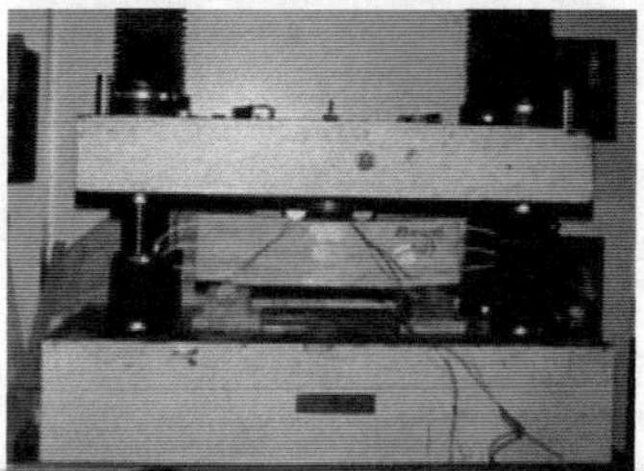

LAND | DEEPWATER | AEROSPACE | **astrotechnology.com**

astro technology

HARDENED SENSOR STATIONS

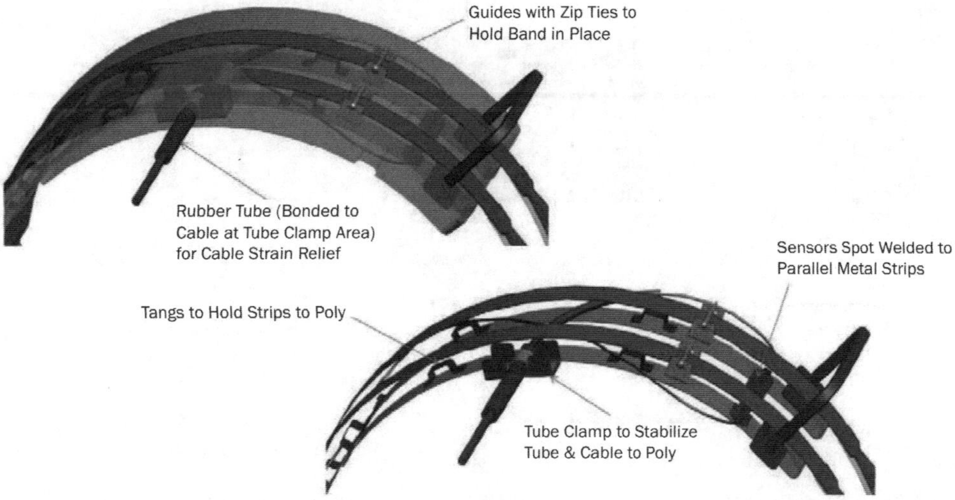

LAND | DEEPWATER | AEROSPACE | **astrotechnology.com**

OBSERVING WAVE ACTION

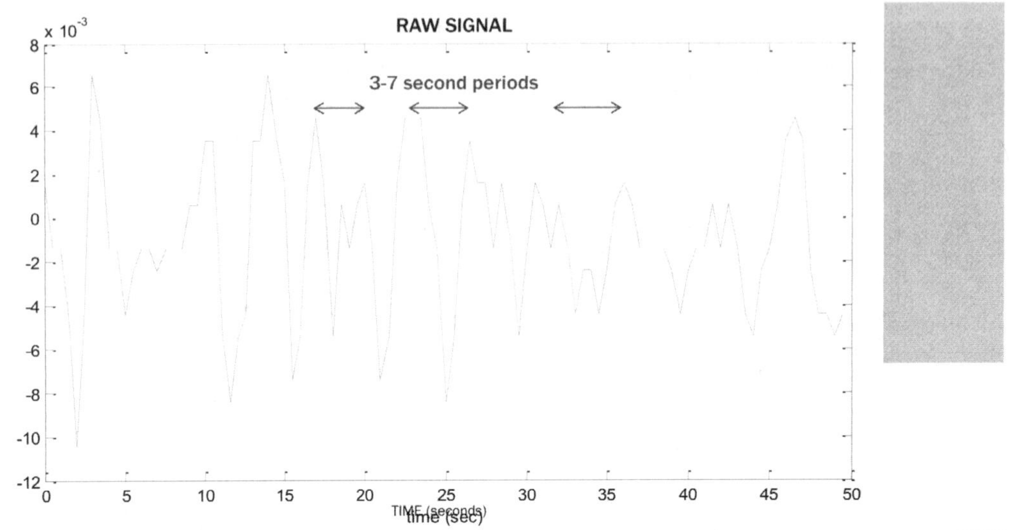

LAND | DEEPWATER | AEROSPACE | **astrotechnology.com**

AMPLITUDE SPECTRUM

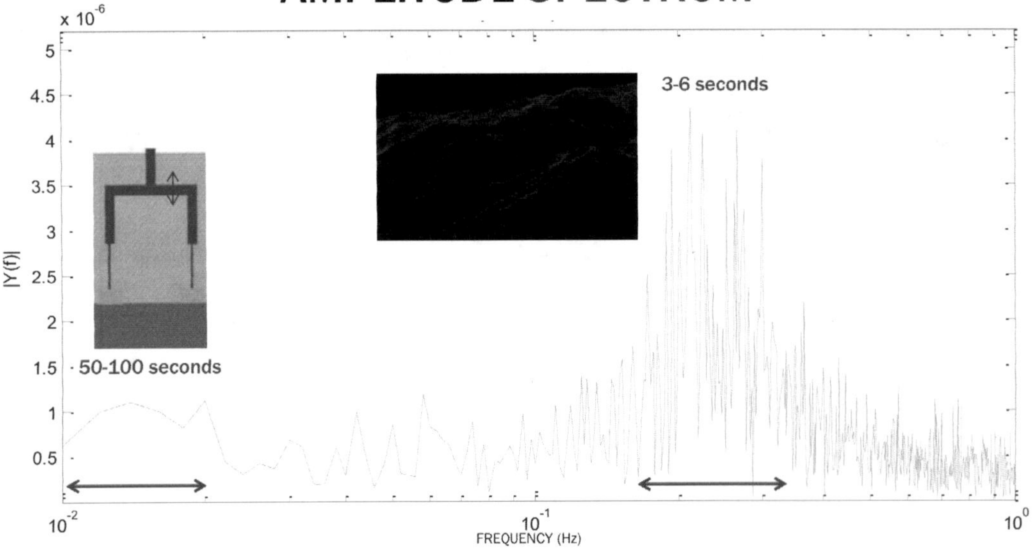

LAND | DEEPWATER | AEROSPACE | **astrotechnology.com**

352754 Characterization of Gas Hydrates in Turbulent Two-Phase Flow Environment Using CFD: Tracking Deposition, Aggregation and Dissociation Effects

Abstract:

Prediction and mitigation of gas hydrates continue to dominate the research in the oil and gas community owing to their ability to substantiate severe production losses and risk issues in many gas and oil fields. As the industry looks towards more reliable modeling for risk assessment and flow assurance, numerical tools continue to expand as a major asset in the research and development for the prediction and mitigation of gas hydrates.

In this paper, we establish a numerical framework to model the formation and transport of hydrate particles in an oil-water pipeline with finite concentration of dissolved methane in a horizontal flow line. Multiphase flow of dispersed water and gas in an oil dominated pipeline is captured using an Eulerian multi-fluid modeling method while accounting for the exchange of mass, momentum and energy between different phases. In particular, formation and transport of hydrate particles generated as a result of dispersed water droplets interaction with the dissolved methane gas, under thermodynamically favorable condition, is tracked using a population balance model. The nucleation process is modeled as a heterogeneous reaction, with intrinsic kinetics, between the water and gas phases leading to finite size hydrate particles. Subsequent growth, aggregation and deposition of the hydrate particle and its influence on flow modulation such as flow jamming and blockage are captured in an extensive fashion. Based on the incoming gas composition, the hydrate formation threshold temperature is established. The effective viscosity of the oil-water is modeled

The current study explores the effect of different incoming gas compositions (effective hydrate formation temperature), water cuts and pipe sizing on the hydrate formation rates and the resulting pressure gradients across the flow line under consideration. Contour plots of hydrate fraction and their size distribution along with velocity and pressure distribution provide a detailed understanding of the flow physics associated with the hydrate formation process.

349380 Assessing the Performance of a Kinetic Hydrate Inhibitor Using a Crystal Hydrate Inhibition (CGI) Method for a Gas and Gas-Condensate System

Abstract:

Gas hydrate formation in oil and gas pipelines is a complex process spanning space and time dimensions across multiple scales. At the micro/meso-scopic scales, (heterogeneous) nucleation and growth occurs. During nucleation, the first hydrate proto-nuclei (clusters) are formed, which, after some time 'induction time', can grow to produce detectable macroscopic crystals. Then, hydrate crystals may agglomerate, potentially forming a plug in the pipeline.

One of the methods to manage hydrate formation is the injection of Kinetic Hydrate Inhibitors (KHIs). The accepted understanding is that KHIs slow down or interfere with nucleation, hence extending the induction time. KHIs are normally evaluated by measuring induction times in the laboratory. However, quantifying induction times is problematic as nucleation is inherently stochastic, meaning measurements are often poorly repeatable with such studies time-consuming. Furthermore, it is often mistakenly believed that when a hydrate crystal has been detected in a KHI inhibited system, the KHI has 'failed'.

Recent research has proven that KHIs are not only nucleation delayers but also powerful crystal growth inhibitors. In view of this, a Crystal Growth Inhibition (CGI) method [1,2] for the reliable evaluation of KHIs has been previously developed. This method shows that KHIs induce a number of well-defined hydrate crystal growth inhibition (CGI) regions with different growth rates as function of subcooling. This new methodology provides a useful decision-making tool in order to manage the uncertainty of KHI performance during the different phases of real field production operations [3].

In this contribution, we present the results of an extensive experimental programme utilising the CGI method for the qualification of a commercial KHI that will be implemented in a greenfield development in the North Sea.

[1] Anderson et al. (2011) Development of a crystal growth inhibition based method for the evaluation of kinetic hydrate inhibitors, 7th Intl Conference on Gas Hydrates

[2] Anderson et al. (2011) Application of a new crystal growth inhibition based KHI evaluation method to commercial formulation assessment, 7th Intl Conference on Gas Hydrates

[3] Luna-Ortiz et al. (2013) Fast-track flow assurance design for kinetic hydrate inhibitors without laboratory testing: A case study, in 16th Intl Conference on Multiphase Production Technology

PROSYS™ Maximize Your Plant.

Operator Errors and What Can be Done to Minimize

Global Congress on Process Safety

March 31, 2014

10TH GLOBAL CONGRESS ON PROCESS SAFETY

PROSYS™ Maximize Your Plant.™

10TH GLOBAL CONGRESS ON PROCESS SAFETY

Presenter

Tom Nolan

- **Graduated from Ohio University with B.S. in Chemical Engineering**
- **24 years experience in chemical process industry in a variety of roles**
- **MAIC and DFSS Six Sigma Black Belt**

PROSYS Maximize Your Plant.™

The Cost of Errors

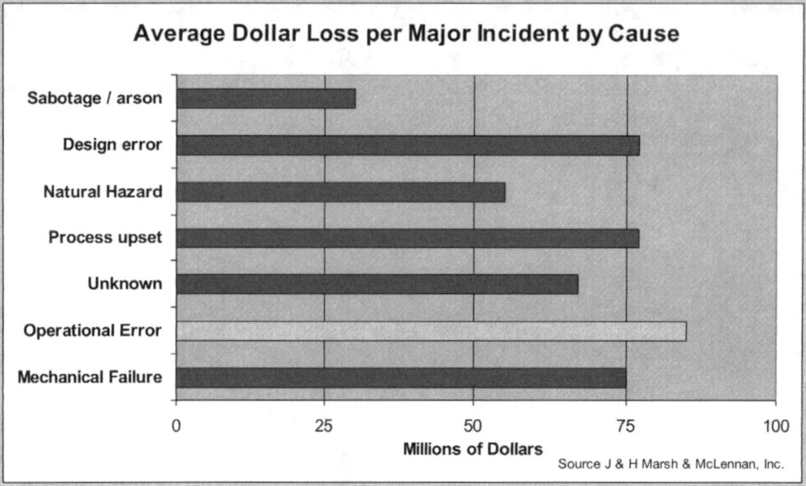

PROSYS Maximize Your Plant.™

The Cost of Operator Errors

- **ASM estimates total loss due to operator error is $8B per year**
- **Chemical Safety Topical Committee - average of one chemical incident per day - cost of over $2 million per incident to <u>comply with requirements</u>**
- **Errors cause 42% of unscheduled shutdowns**
- **70% of process incidents occur during start-up or shutdown**

PROSYS Maximize Your Plant.™

10th GLOBAL CONGRESS ON PROCESS SAFETY

Reasons for Errors

- Lack of Skill
- Lack of Knowledge
- Carelessness
- System Design
  - Operator set up to make errors by inappropriate design – built in errors
  - Errors that are predictable are preventable by better design

PROSYS Maximize Your Plant.™

10th GLOBAL CONGRESS ON PROCESS SAFETY

Focus of this Discussion

- System Design
  - Alarm Management
  - Operator Graphics
  - Operator set up to make errors by inappropriate design – built in errors
  - Errors that are predictable are preventable by better design

Alarm Floods

In a number of industrial incidents, alarm floods were identified as a significant contributing cause to the incident...

As found by EEMUA in 1999 and CSB

Alarm Flood defined by ISA 18.2 as -
"10 or more annunciated alarms in any 10 minute period per operator"

Why do Alarm Floods Occur?

One reason is not providing dynamic alarm management

- **Alarms need to indicate abnormal situations that require operator action**
- **Processes do not operate in one state**
- **What is normal vs abnormal changes with state**
- **Alarms are typically configured for run therefore many alarms are triggered upon a Change of State: Run to Shutdown**
- **Many of the alarms are not applicable or actionable for the new state impleading the operators ability to act quickly on what is important**

PROSYS Maximize Your Plant.™

10th GLOBAL CONGRESS ON PROCESS SAFETY

What is Impacted by Alarm Floods?

- **Product quality**
- **Operability or profitability of the process**
- **Loss of equipment**
- **Loss of containment – environmental releases**
- **Injury and loss of life in plant or community**

PROSYS Maximize Your Plant.™

10th GLOBAL CONGRESS ON PROCESS SAFETY

What Makes Alarm Floods so Dangerous?

Can be a problem for three reasons:

- **A deluge of alarms can cause critical alarms to be missed**
- **Floods can be a significant distraction when dealing with process upsets**
- **Can be an indicator of larger systemic safety issues**

PROSYS Maximize Your Plant™

10th GLOBAL CONGRESS ON PROCESS SAFETY

Impacting Alarm Management Design

- Alarm rationalization is not a process to eliminate alarms - its about quality
- Good rationalization will add alarms when appropriate
- Typical (Static) rationalization is only for run mode
- Dynamic rationalization considers all plant modes

PROSYS Maximize Your Plant™

10th GLOBAL CONGRESS ON PROCESS SAFETY

Dynamic Alarm Management

- Dynamic rationalization does everything a static rationalization does plus asks "when" for each alarm
- Requires dynamic software to make changes based upon operating mode of the plant
- Eliminates redundant alarms and lowers operator loading during transitions
- Only alarms what is abnormal and actionable for the given state

PROSYS™ Maximize Your Plant.™

10th GLOBAL CONGRESS ON PROCESS SAFETY

Actual Performance Metrics vs ISA 18.2

Type	Avg Alarm Rate/hr	Low/Hi of Avg Rate/hr	Peak Alarm Rate/hr	Low/Hi of Peak Rate/hr	% Time in Flood	Standing Alarms >24 hrs
Before Rationalization	30	Low=6.2 Hi=61	638	Low=152 Hi=2402	17.2%	9
ISA 18.2 Target Metrics	6	---	<60	---	<1%	<5
After Dynamic Alarm Management	2	Low=0.09 Hi=2.5	25	Low=13 Hi=42	0.25%	4

PROSYS™ Maximize Your Plant.™

10th GLOBAL CONGRESS ON PROCESS SAFETY

Results After Dynamic Alarm Management

Annunciated Alarms Per Hour (Monthly) **Area: G1-G5-G6** **June 2013**

Average Counts | Peak Counts | Total Counts

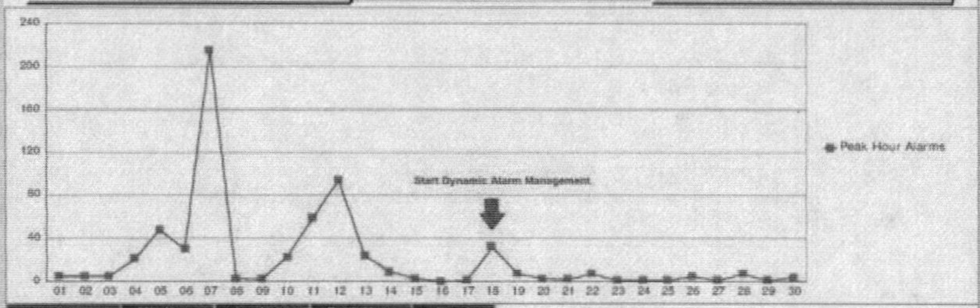

Day	Min Hour Alarms	Avg Hour Alarms	Peak Hour Alarms	Total Alarms
01-Jun-2013	0	0.3	5	6
02-Jun-2013	0	0.4	5	9
03-Jun-2013	0	0.5	5	13
04-Jun-2013	0	1.1	21	27
05-Jun-2013	0	13.3	48	319
06-Jun-2013	0	3.6	30	86
07-Jun-2013	0	11.9	215	285
08-Jun-2013	0	0.4	3	10
09-Jun-2013	0	0.3	2	7
10-Jun-2013	0	1.5	23	35
11-Jun-2013	0	4.7	59	112
12-Jun-2013	0	11.9	94	285
13-Jun-2013	0	3.8	24	90
14-Jun-2013	0	1.3	9	32
15-Jun-2013	0	0.3	3	8

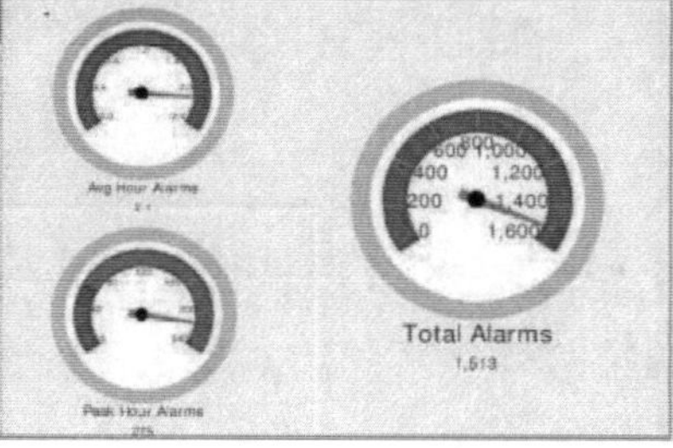

PROSYS Maximize Your Plant.™

10TH GLOBAL CONGRESS ON PROCESS SAFETY

Operator Graphics – The Risk of Changing Focus

- **Distraction of changing focus from process graphic to faceplate window**
- **Added workload from managing multiple open windows**
- **Increased probability of errors when changing values for un-intended tag with multiple faceplates open**
- **All of above become multiplied and more complex when process is transitioning from one state to another**

PROSYS Maximize Your Plant.™

10TH GLOBAL CONGRESS ON PROCESS SAFETY

Faceplate Operation

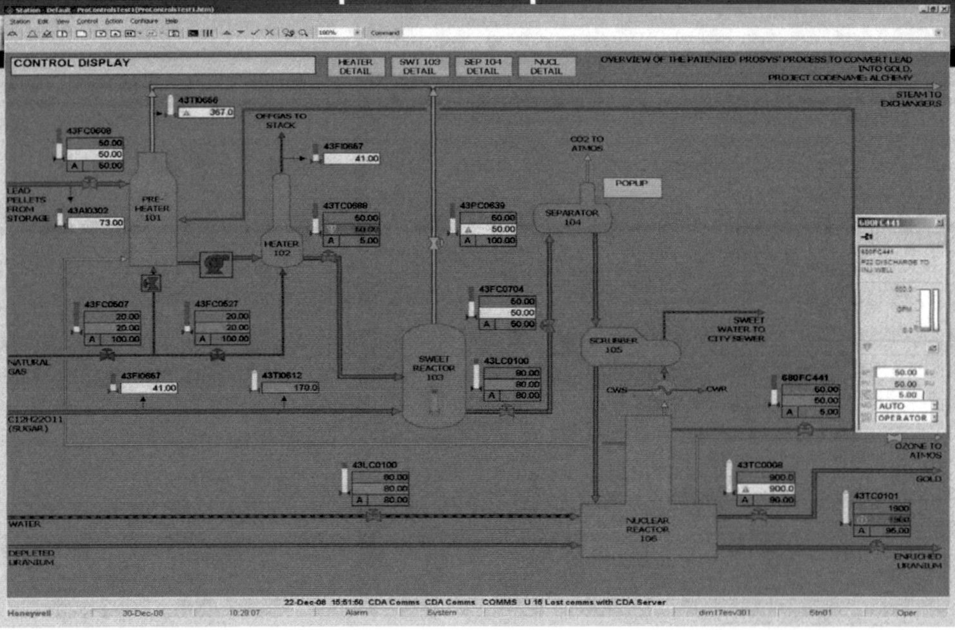

PROSYS Maximize Your Plant™

Potential for Errors

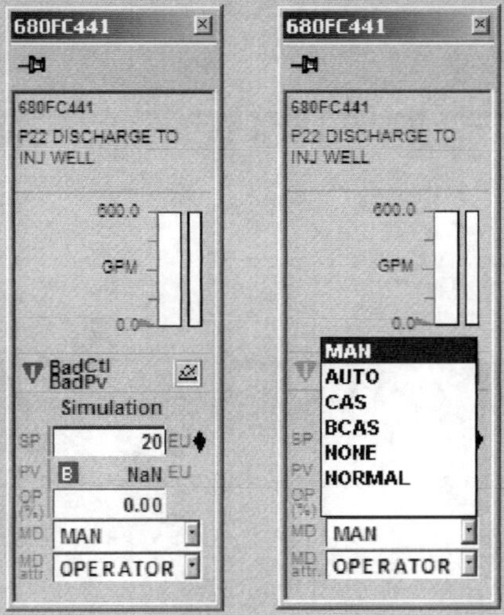

Faceplate Design -can introduce additional errors

- Mode drop down list covers SP, PV and OP values
- Operator may select wrong Mode if SP, PV and OP values are hidden from view
- Mode list offers more options than necessary

PROSYS Maximize Your Plant™

Direct Entry Operation

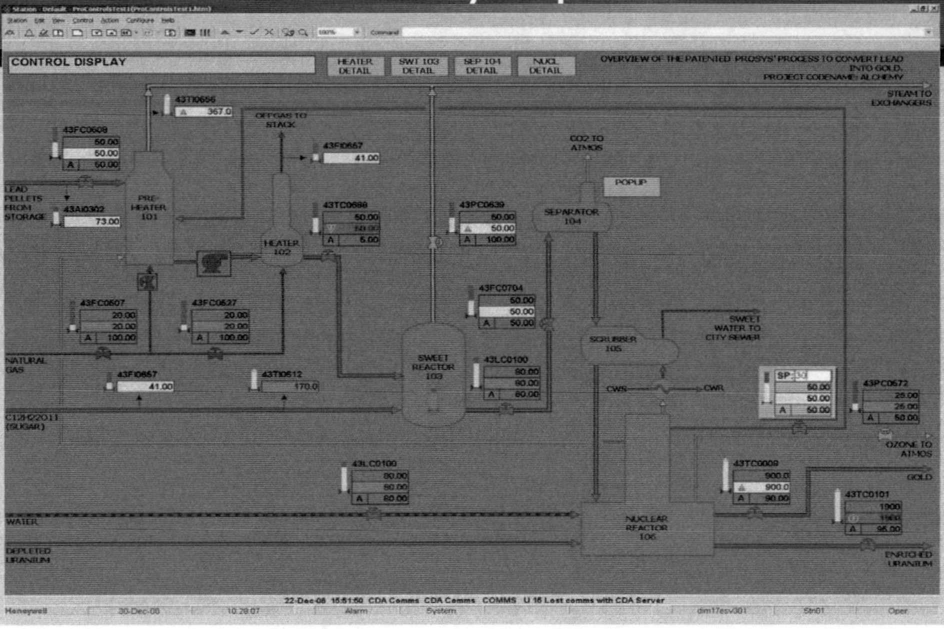

PROSYS Maximize Your Plant™

10th GLOBAL CONGRESS ON PROCESS SAFETY

Direct Entry Fields

Values can be entered by:

- Select point
- Typing value via keypad
- Pressing enter

Key Factors:

- By maintaining focus of points after value change, it improves safety for input errors, multiple clicking functions and is more time efficient
- Easily recognizable operator enterable fields

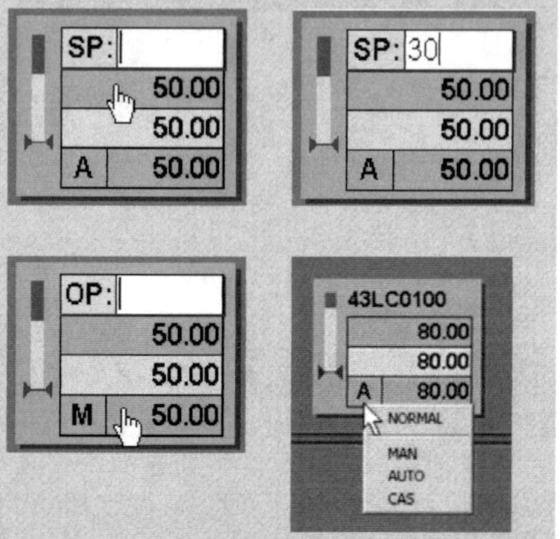

PROSYS Maximize Your Plant™

10th GLOBAL CONGRESS ON PROCESS SAFETY

Evaluation of Methods

Comparison of Faceplate to Direct Entry

- **Keystroke Level Analysis**
- **Operator Loading Analysis**
- **Risk Analysis**

PROSYS Maximize Your Plant.™

10th GLOBAL CONGRESS ON PROCESS SAFETY

Keystroke Level Model

Symbol	Time (s)	Description
K	0.28	Keystroke
P	1.1	Point to object
BB	0.2	Click on object
H	0.4	Home hands on keyboard or mouse
M	1.2+	Mental act or routine thinking

PROSYS Maximize Your Plant.™

10th GLOBAL CONGRESS ON PROCESS SAFETY

Estimate Execution Time Analysis

- **Evaluates execution time by an operator comparing faceplates to direct entry for SP/OP and Mode Changes**

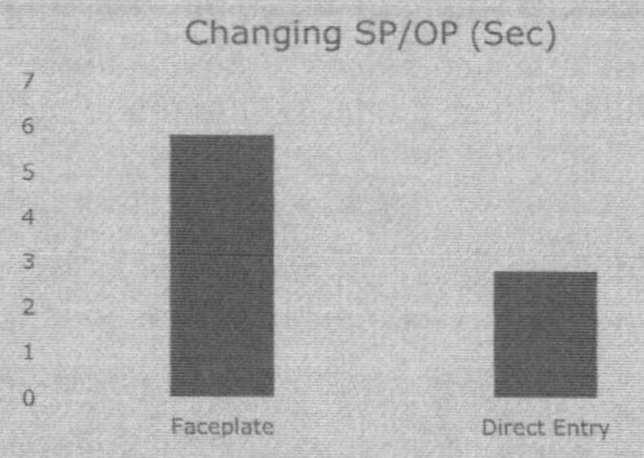

- **50 % Reduction**

PROSYS Maximize Your Plant.™

10th GLOBAL CONGRESS ON PROCESS SAFETY

Estimated Operator Loading Analysis

- **Evaluates Mental and Physical Load on Operator**
  - Tasks that require very little thought are removed – such as (BB), (H)

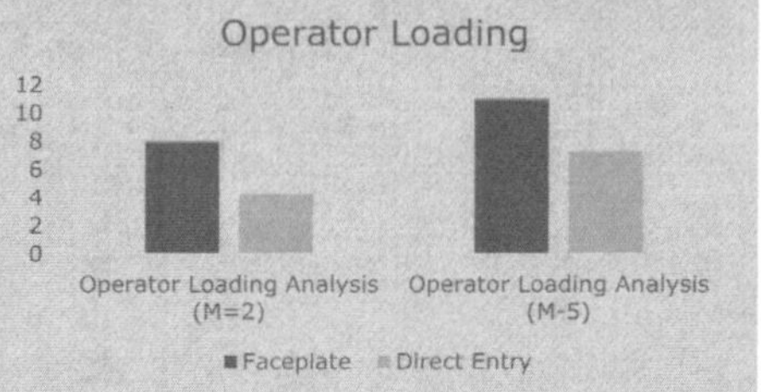

  - Mental and Physical tasks are weighted based on the amount of fatigue they cause – assigned a weight of M=2.0 to 5.0

PROSYS Maximize Your Plant.™

10th GLOBAL CONGRESS ON PROCESS SAFETY

Risk Analysis

- **Identifies actions where operator entry errors can occur**

PROSYS Maximize Your Plant.™

10th GLOBAL CONGRESS ON PROCESS SAFETY

Risk Analysis

Faceplate operation KLM for changing SP/OP	Potential Error
Move to shape (P)	Select wrong point
Click on shape (BB)	
Move to faceplate (P)	Point in faceplate is previous point
Click on SP/OP field (BB)	Change wrong parameter
Move hands to keyboard (H)	
Type in value and press enter (4K)	Mistype value and press enter
Move hands to mouse (H)	
Move to close faceplate (P)	
Click to close faceplate (BB)	
Potential error count	**4**

PROSYS Maximize Your Plant.™

10th GLOBAL CONGRESS ON PROCESS SAFETY

Risk Analysis

Direct Entry operation KLM for changing SP/OP	Potential Error
Move to SP/OP (P)	Select wrong parameter or point
Click on SP/OP field (BB)	
Move hands to keyboard (H)	
Type in value and press enter (4K)	Mistype value and press enter
Potential error count	**2**

PROSYS™ Maximize Your Plant™

10th Global Congress on Process Safety

Direct Entry Vs Faceplate

Summary of Benefits

- **Execution Time SP/OP Changes - 51.5% less**
- **Execution Time for Mode Changes – 50% less**
- **Operator Loading – 33 % to 47% less**
- **Risk of Errors – 50% less**

PROSYS™ Maximize Your Plant™

10th Global Congress on Process Safety

Conclusion

- **Many industrial errors are a result of operators using control systems with flawed designs**
  - Poor design and performance by alarm management distracts operators and/or occludes critical alarms
  - Operator graphics with multiple open faceplates can cause a change intended for one controller to be entered into the faceplate of another
- **The two mechanisms listed above are predictable and therefore preventable through better design available today**

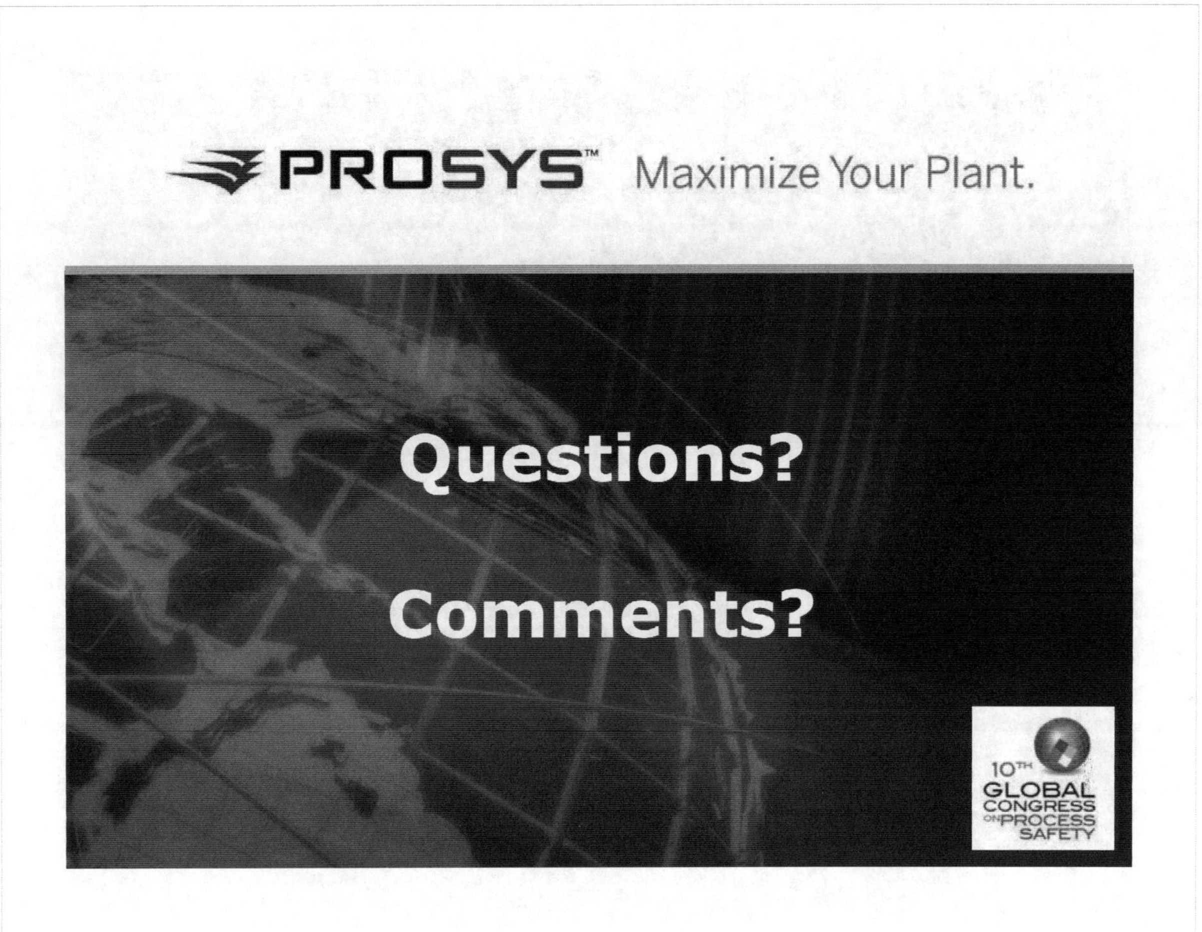
PROSYS™ Maximize Your Plant.
Questions?
Comments?
10TH GLOBAL CONGRESS ON PROCESS SAFETY

Problem Statement

visium KMS
Process Safety Management Solution

- Any organisation can improve by leveraging knowledge and lessons learned.
- Knowledge can be externally or hidden locally, in a global enterprise, where lessons learned are difficult to share beyond their initial impact.
- There is so much data, information and knowledge (opportunities) coming at your company that can make you stronger or make you look like your not paying attention.
- Not paying attention or managing the available knowledge, can sink your ship… nobody wants that!

Expert Insight Delivered

2

OSyS
Part of the Rolls-Royce Group

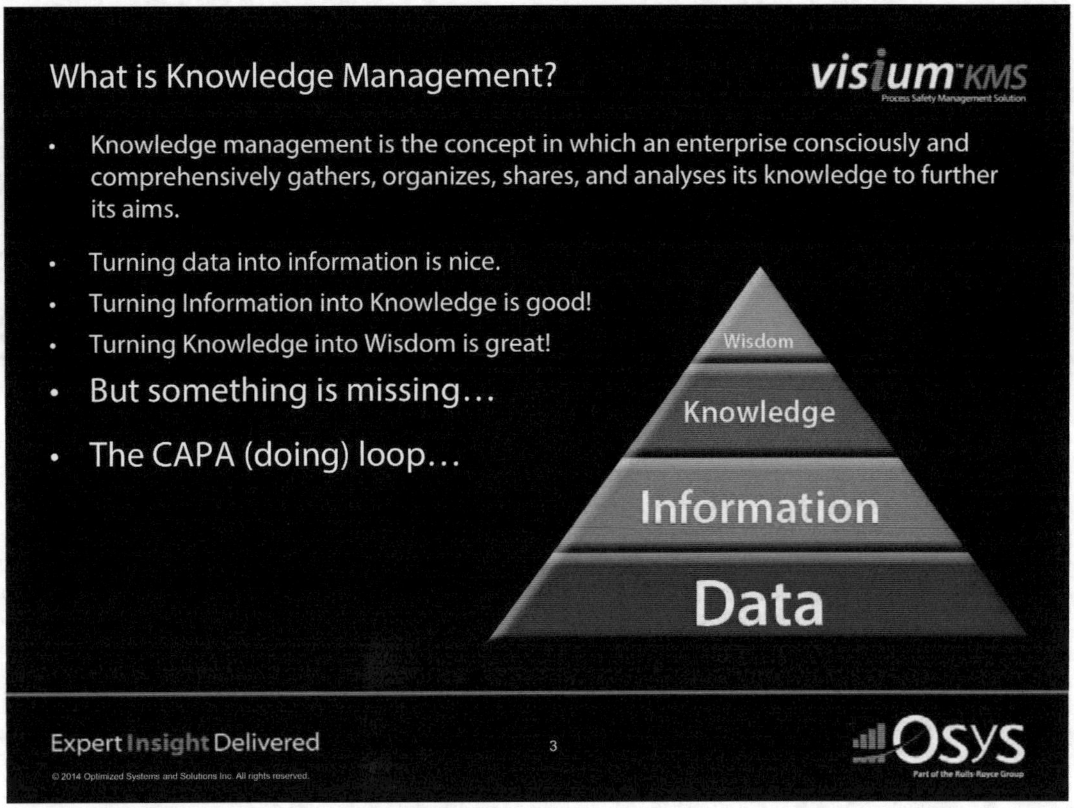
What is Knowledge Management?
visium KMS
Process Safety Management Solution
• Knowledge management is the concept in which an enterprise consciously and comprehensively gathers, organizes, shares, and analyses its knowledge to further its aims.
• Turning data into information is nice.
• Turning Information into Knowledge is good!
• Turning Knowledge into Wisdom is great!
• But something is missing…
• The CAPA (doing) loop…
Wisdom
Knowledge
Information
Data
Expert Insight Delivered
3
© 2014 Optimized Systems and Solutions Inc. All rights reserved.
Osys
Part of the Rolls-Royce Group

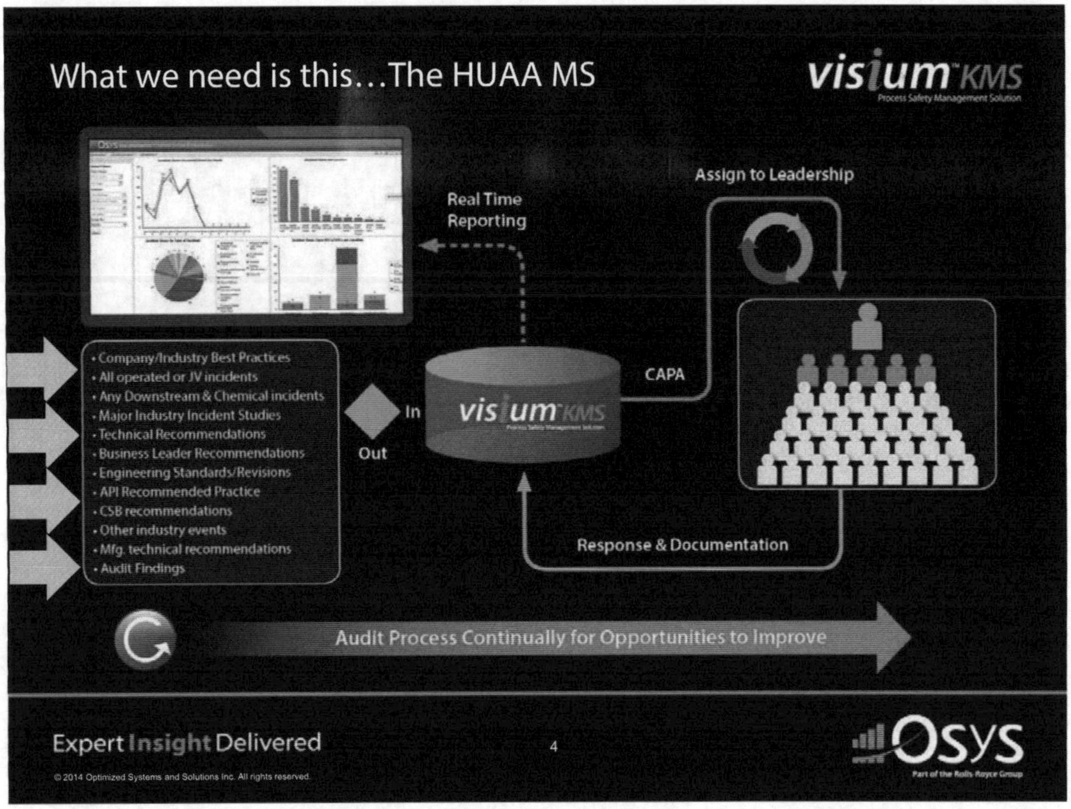
What we need is this…The HUAA MS
visium KMS
Process Safety Management Solution
Real Time Reporting
Assign to Leadership
• Company/Industry Best Practices
• All operated or JV incidents
• Any Downstream & Chemical incidents
• Major Industry Incident Studies
• Technical Recommendations
• Business Leader Recommendations
• Engineering Standards/Revisions
• API Recommended Practice
• CSB recommendations
• Other industry events
• Mfg. technical recommendations
• Audit Findings
In
Out
visium KMS
CAPA
Response & Documentation
Audit Process Continually for Opportunities to Improve
Expert Insight Delivered
4
© 2014 Optimized Systems and Solutions Inc. All rights reserved.
Osys
Part of the Rolls-Royce Group

What is the HUAA Management System…

visium KMS
Process Safety Management Solution

HUAA – Heard, Understood, Acknowledged, Actioned.

1. Opportunities identified (heard) by your experts,
2. Captured by an electronic system,
3. Expertly Reviewed (understood),
4. Accepted or rejected (acknowledged) and
5. Assigned (actioned) to Sr. Managers for resolution and tracked to closure.

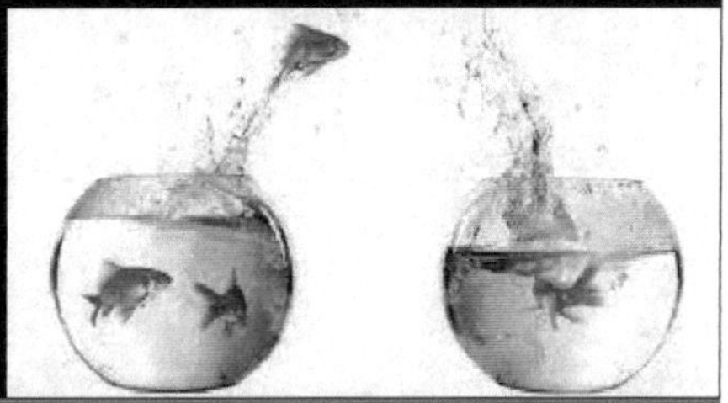

Expert Insight Delivered   5

OSyS
Part of the Rolls-Royce Group

HUAA – Management Systems - Detail

visium KMS
Process Safety Management Solution

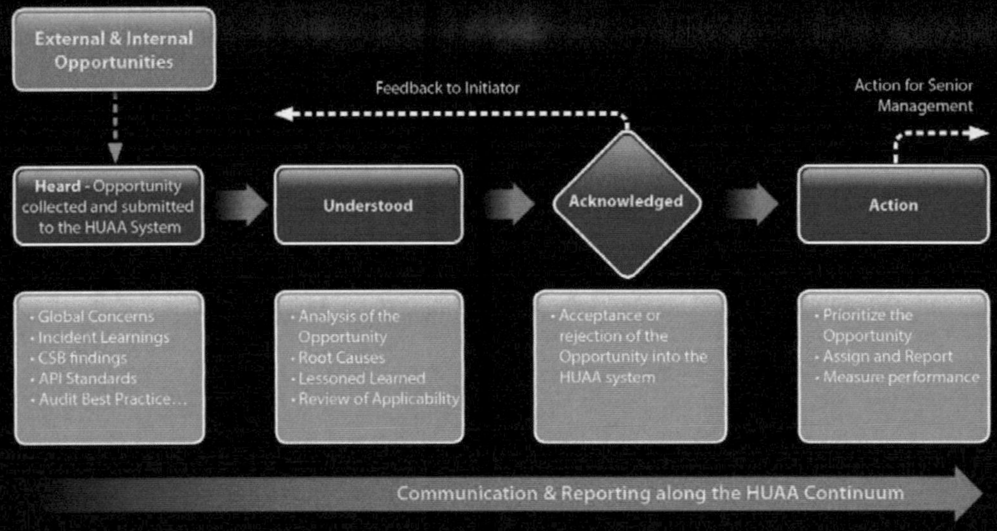

Expert Insight Delivered   6

OSyS
Part of the Rolls-Royce Group

HUAA Results

visium KMS
Process Safety Management Solution

- Efficient Collection - Important Opportunities are identified and collected from many sources (internal and external), and entered into the HUAA management system.
- Expert Vetting - These opportunities are vetted, parsed and resolved for inclusion or exclusion; documenting & communicating both.
- Accountable Assignment - Included opportunities are systematically assigned to Sr. Leaders and managed across the enterprise at SVP and C levels.
- Action - Improvement will be assured through the accountable actions of these leaders.
- Results - The organization will gain competitive advantage by active listening, learning from their experience, as well as the experience of others, and by taking ACTION!

Expert Insight Delivered 7

OSYS
Part of the Rolls-Royce Group

Sharing is not enough to keep your ship afloat!

visium KMS
Process Safety Management Solution

Expert Insight Delivered 8

OSYS
Part of the Rolls-Royce Group

352631 Autocad - Smart Grid – Agriculture - Architecture and the United States Government

Abstract:

The computational methods, process control, process design, process safety, public policy of the smart chip for agriculture industry, and architecture industry in the United States utilizing the computational methods of AutoCAD in the structural development stages of the smart grid for device/component/equipment and geographical development and public policy for grant funding.

352747 Production, Characterization and Catalytic Studies of Biobased Carbon Materials

Abstract:

Carbon-based nano materials were produced from wood chips, lignin and bio-char. The influence of reaction conditions on the production of nano-carbons was investigated. Carbon-based nanomaterials were characterized using scanning electron microscopy (SEM), transmission electron microscopy (TEM), high resolution transmission electron microscopy (HRTEM), and X-ray diffraction (XRD). The catalytic performance of these nanomaterials as syngas cleaning adsorbents and chemical conversion catalysts for bio-fuel processes has been evaluated.

347835 Uncertainty in Sour Gas Viscosities Estimation, What Is the Effect on Your Reservoir Inflow and Tubing Performance

Abstract:

In light of the need to find an applicable, consistent and promising correlation to accurately calculating sour gases viscosity under several reservoir conditions, this study aims to compare the existing correlations and other industry standard methods, check their accuracy and range of applicability for sour/acidic gases. Moreover, investigating the impact of uncertainties in estimating sour gas viscosities on reservoir flow performance relationship (IPR), reserves and resources estimations, tubing performance, pipeline and facilities design, the design of compressors and more importantly the need to find such correlation to be used in further advancements of reservoir simulation.

354052 Chromonic Nanocarriers for Chemotherapeutics: Size Distribution and Control Release Studies

Abstract:

Conventional chemotherapy fails to release the drug in the controlled manner; rather they release the drug instantly due to which re-administration is required as therapeutic index falls. Moreover, they target both the cancer cells and non-cancer cells which make the treatment of the cancer cells highly ineffective due to excessive toxicities. Recently, the nanoparticle mediated drug delivery for chemotherapeutics has gained much attention. The present study reports the use of a novel class of materials – *Chromonics* in cancer drug delivery. Folic acid as a chromonic material has been explored to develop nanocarrier for sustained release of chemotherapeutic drugs *in-vitro*. With the help of a biocompatible polymer as a tool, folate nano-domains were formed from self-assembled liquid crystalline folate solutions. These nano-domains were crosslinked to form a stable nanoparticle system. The effect of different factors like crosslinking agent, relative polymer and folate concentration, on the size distribution was investigated in detail. Based on the observations, we have reported that not only the synthesis but also the size of the developed nanoparticles can be controlled by varying the folate: HPMC weight ratios. The present study then explored the release of drugs from these nanocarriers. We have investigated the release of folic acid as a model drug from these nanoparticles as well as release of methotrexate and cytarabine as an anticancer drug for the period of 60 days. It was observed that release of drug was nanoparticle's size-dependent, release medium dependent, pH of the release medium dependent, crosslinking agent and its concentration dependent. The significant part of the current study is the synthesis of chromonic nanoparticles, how the size can be controlled, how controlled release can be obtained as well as how the release profiles can be modified depending on the dosage forms. Therefore, we can say that folate nanoparticles can be a potent nanocarrier for the sustained release of cancer drugs.

Double Containment Piping Systems: A Fail-Safe, Leak-Free Solution

IPEX
Committed to Excellence

Darin Johnson, P.Eng.
Applications Engineer
IPEX USA LLC

AIChE – 2014 Spring Meeting
New Orleans, LA

Monday, March 31, 2014

Why Double Containment Piping Systems?

IPEX
Committed to Excellence

- December 1988 – law change within the ***Resource, Conversation, and Recovery Act*** in the United States regarding underground piping.
- U.S. Environmental Protection Agency (EPA) requirement: Code of Federal Regulations (CFR), Title 40, Part 280.
  - Petroleum Products
  - Hazardous Substances (over 3200, CFR Title 40 Part 302.4)
  - Hazardous Wastes

Why Double Containment Piping Systems?

- Buried tank - all buried piping connected to it must be double contained.
- Above ground tank – associated buried piping consists of 10% or more of the tank volume, it also must be double contained.

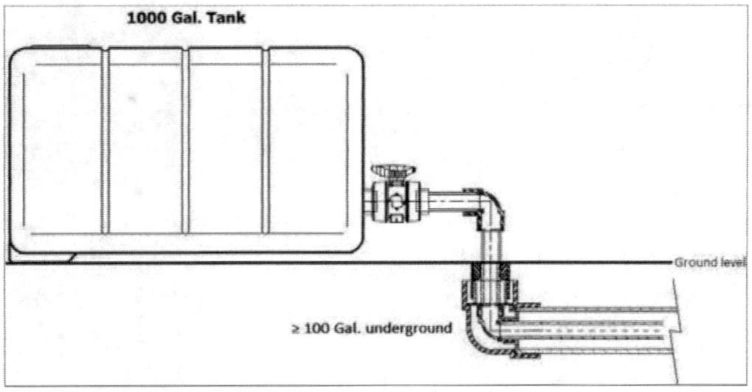

Why Double Containment Piping Systems?

- If the piping is less than 10% of the volume of the aboveground tank, and you choose not to double contain the buried piping, ***responsibility for any clean-up or remediation resulting from a release will still fall on the Client.***

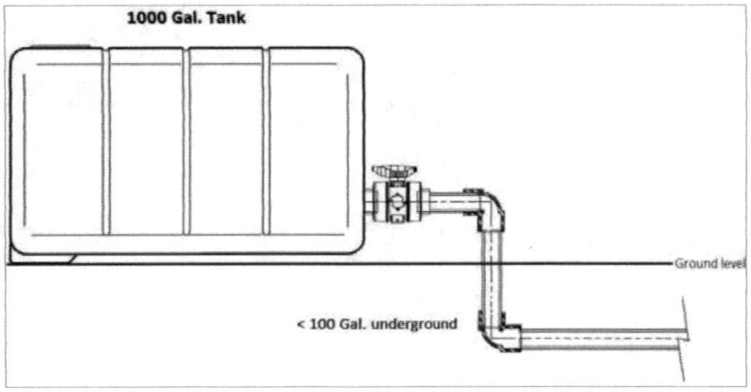

Why Double Containment Piping Systems?

- **Above Ground Piping Systems**
  - Different jurisdictions have to ability to require chemicals be double contained.

Nomenclature of Double Containment Piping

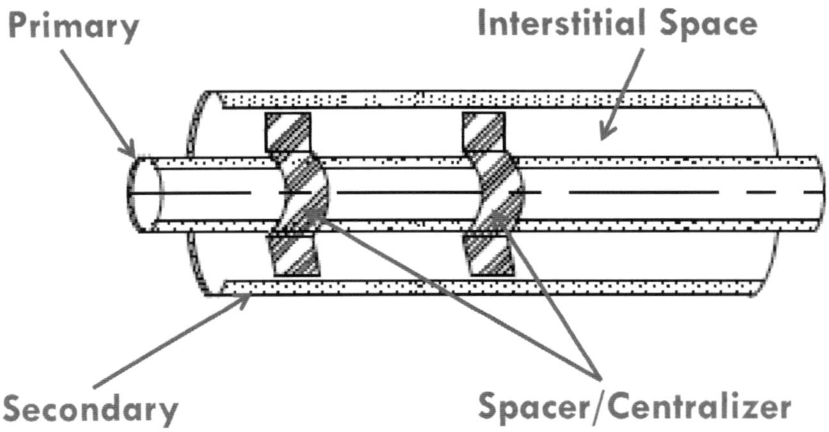

Double Containment Piping Systems: Material Combinations

SYSTEM Carrier	Containment	Carrier SIZE inches
PVC 80	PVC 80	1/2 - 12
PVC 80	PVC 40	1/2 - 12
PVC 40	PVC 40	1/2 - 12
PVC DWV	PVC DWV	1-1/2 - 12
PVC 80	CS SCH 10	1/2 - 12
CPVC 80	CPVC 80	1/2 - 8
CPVC 80	PVC 80	1/2 - 12
CPVC 80	PVC 40	1/2 - 12
CPVC 80	CS SCH 10	1/2 - 12
NAT PP 80	PVC 80	1/2 - 2
NAT PP 80	PVC 40	1/2 - 2
RED PVDF 80	CPVC 80	1/2 - 2
RED PVDF 80	PVC 80	1/2 - 2
NAT PVDF 80	PVC 80	1/2 - 2
ENFUSION PP	PVC DWV	1/2 - 6
S40 A53B CS BW	PVC 80	1/2 - 8
S40 A106B CS BW	PVC 80	1/2 - 8

SYSTEM Carrier	Containment	Carrier SIZE inches
S40 A53 BLK SW	FRP	1/2 - 2
S40 A53 BLK SW	PVC 80	1/2 - 2
S80 A53B CS SW	PVC 80	1/2 - 2
S10 T304L SS BW	PVC 80	1/2 - 8
S10 T304L SS BW	PVC 40	1/2 - 8
S10 TO304L SS BW	CPVC 80	1/2 - 8
S10 T304L SS BW	FRP	1/2 - 12
S10 T304L SS BW	S10 T304L	1/2 - 12
S40 T316L SS BW	PVC 80	1/2 - 8
S40 T316L SS BW	CPVC 80	1/2 - 8
S40 T316L SS BW	FRP	1/2 - 12
S40 T304L SS SW	PVC 80	1/2 - 2
S40 T304L SS SW	PVC 40	1/2 - 2
S40 T304L SS SW	CPVC 80	1/2 - 2
S40 T304L SS SW	FRP	1/2 - 2
S10 T316L SS SW	PVC 80	1/2 - 2
S10 T316L SS SW	CPVC 80	1/2 - 2

Double Containment Piping Systems: Design Considerations

- **Material Selection**
  - Primary Pipe
    - Chemical Compatibility
    - Pressure Capabilities
    - Thermal Properties
  - Secondary Pipe
    - Chemical Compatibility
    - Loading (Static & Live) for buried applications
    - Hanging requirements for above ground
    - Leak detection requirements
    - UV requirements

IPEX
Committed to Excellence

Double Containment Piping Systems: Design Considerations

- **Thermal Expansion & Contraction**
  - Two basic thermal relationships may be present:
    - When only one of the two systems experiences thermal expansion/contraction.
    - When both systems experience different magnitudes of thermal expansion/contraction.
  - The Floating System
  - The Restrained System

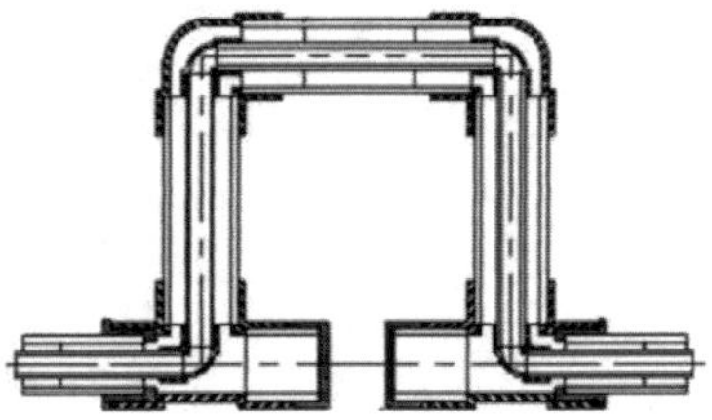

IPEX
Committed to Excellence

Double Containment Piping Systems: Leak Detection Requirements

- EPA requirement: (CFR), Title 40, Part 280
  - Pressurized systems require electronic leak detection that must be checked once per year.
  - Drainage systems (gravity flow) require visual leak detection with monthly manual inspection for a media leak.

IPEX
Committed to Excellence

Double Containment Piping Systems: Leak Detection Requirements

- The following forms of leak detection are acceptable as per regulations
  - Visual (drainage applications only)
  - Electronic low point probes
  - Continuous cable

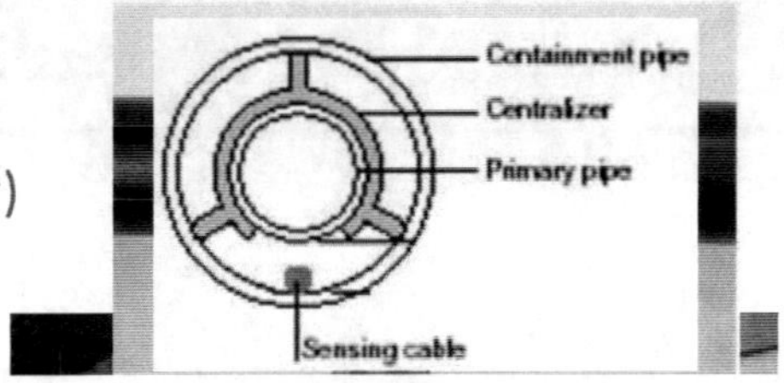

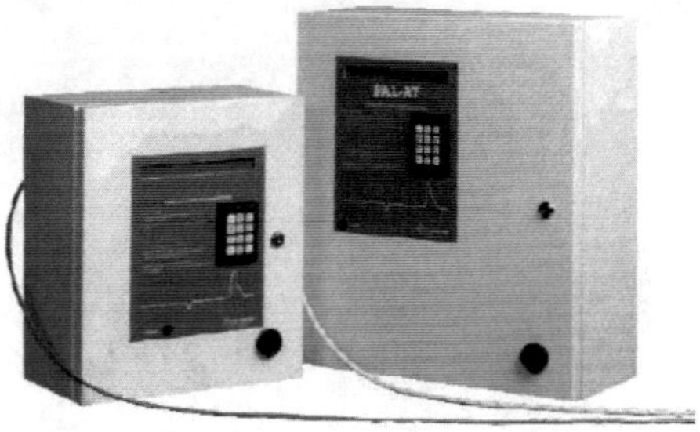

IPEX
Committed to Excellence

Double Containment Piping Systems: Applications

- Underground Applications
- WTP & WWTP
- Oil & Gas
- Hospitals
- Museums
- Food and Beverage
- Pharmaceuticals
- Laboratories
- Data Centers

IPEX
Committed to Excellence

Double Containment Piping Systems: A Fail-Safe, Leak-Free Solution

Questions?

Thank you!

DTP Process

"Now Ready to License"

JGC Corporation
Mitsuo Morita

2014

JGC CORPORATION

MITSUBISHI CHEMICAL

What is DTP Process ?

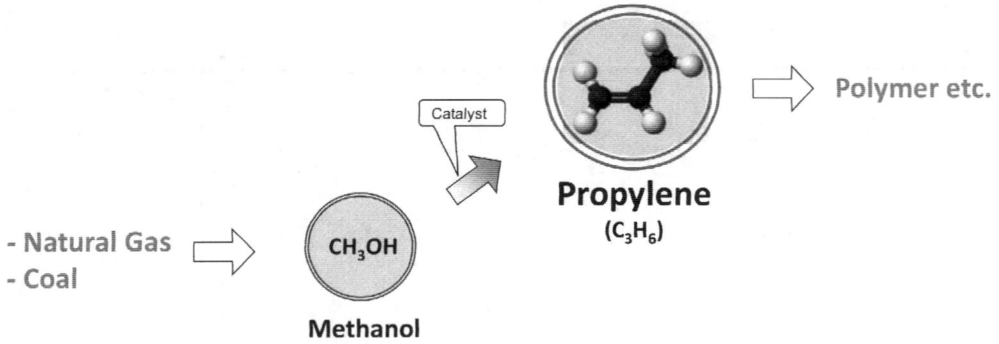

-On-Purpose Propylene Production from Methanol

JGC CORPORATION

- 2 -

MITSUBISHI CHEMICAL

What is DTP Process ?

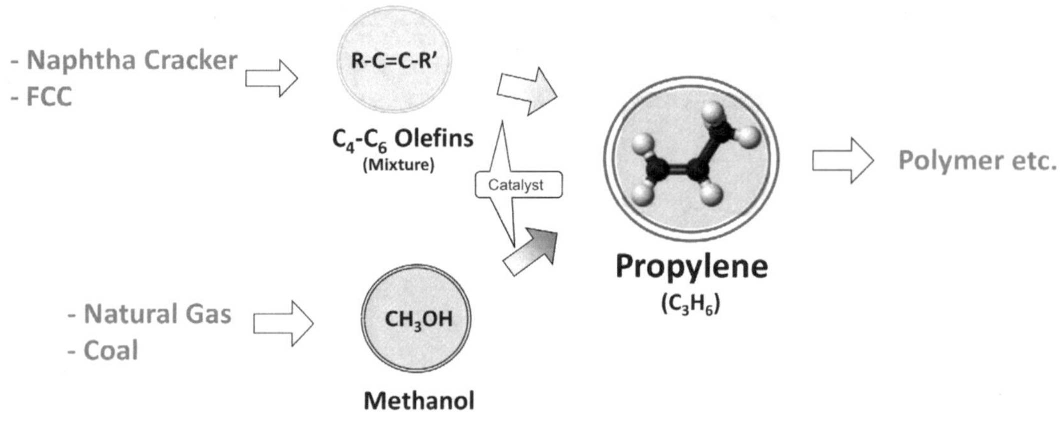

-On-Purpose Propylene Production from Methanol

- Additional Unutilized C_4-C_6 Olefins as Feedstock

JGC CORPORATION - 3 - MITSUBISHI CHEMICAL

Development History and Prospect

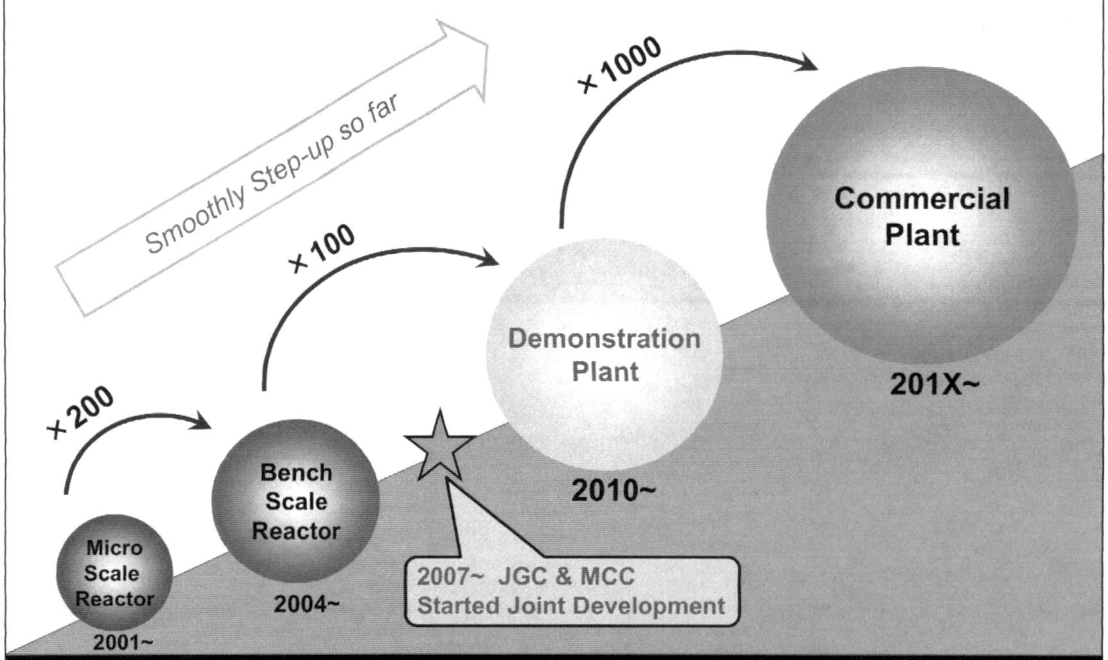

JGC CORPORATION - 4 - MITSUBISHI CHEMICAL

Demonstration Plant Overview

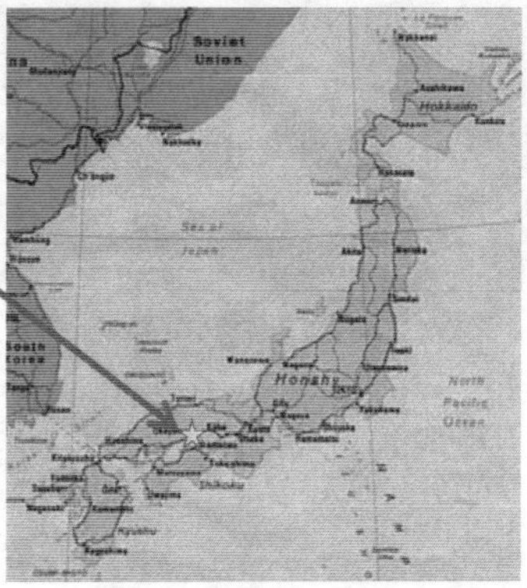

- **Located in Mitsubishi Chemical Mizushima Plant**
- **Successfully Start-up in August 2010**
- **Continuous Operation for Two Years**

Process Flow (Standalone Case)

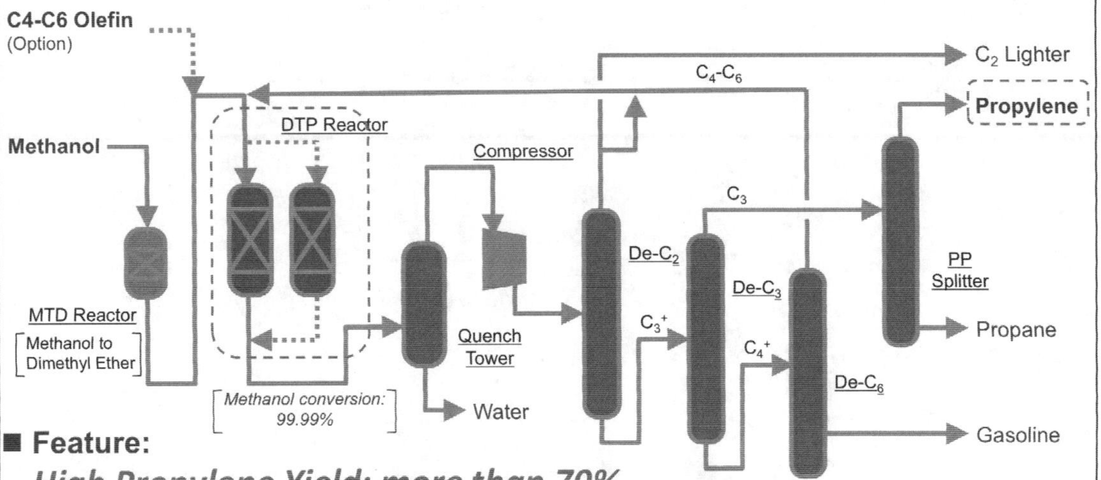

■ **Feature:**

- ***High Propylene Yield: more than 70%***
- ***Applied Existing Proven Technologies***

Process Flow (Integration Case)

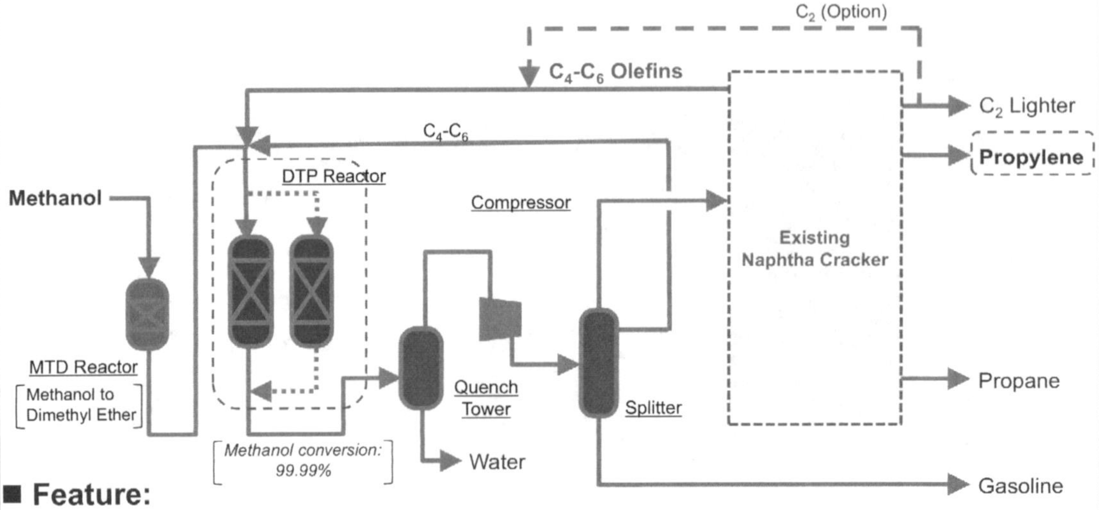

■ **Feature:**

- *High Propylene Yield: more than 70%*
- *Applied Existing Proven Technologies*
- *Low Investment Cost in case of Integration with Naphtha Cracker*

Application

(1) Naphtha Cracker Integration Case (Propylene Capacity: 20KTA – 200KTA)

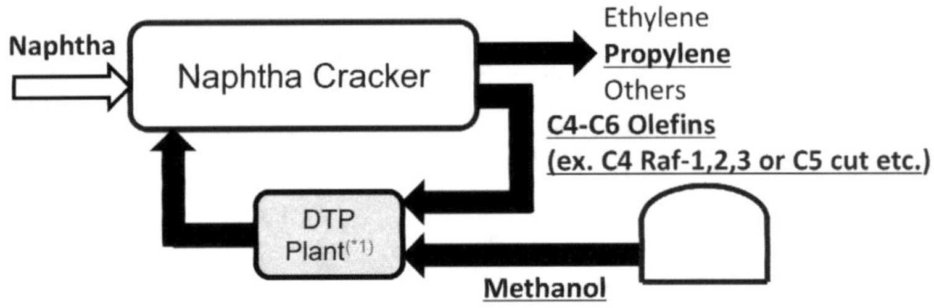

(*1) Utilizing facilities in Naphtha Cracker.

(2) Standalone Case (Propylene Capacity: 250KTA – 500KTA)

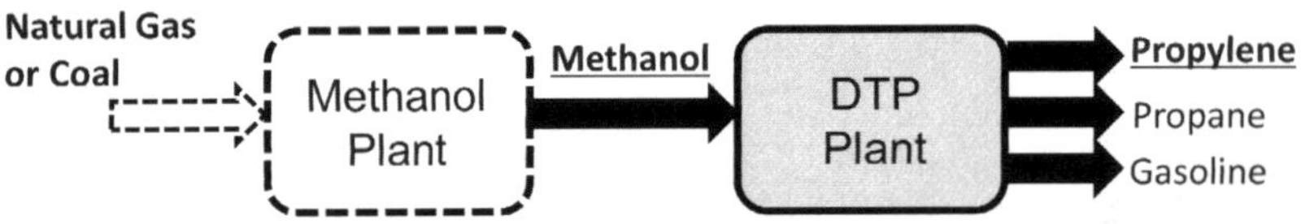

- *Two Cases available depending on Client requirements !!*

DTP is…

- *High Yield Propylene Production Process*
- *Flexible Feed Methanol to Olefin Process*
- *Process that increase P/E Ratio for Cracker*

Thank you very much !!

Contact:
Mitsuo Morita
Market Development Dept.
Global Marketing Division
2-3-1, Minato Mirai, Nishi-ku,
Yokohama 220-6001, Japan

Phone: 81-45-682-8371
FAX: 81-45-682-8183
E-Mail: morita.mitsuo@jgc.co.jp

JGC CORPORATION - 9 - MITSUBISHI CHEMICAL

352533 Self-Cleaning "Bernoulli" Type Filters Used in Onshore and Offshore Applications

Abstract:

Georg Schuenemann GmbH (SAB-BREMEN) has developed a new Filter Type F-480, dubbed the "Compact Giant" for applications in onshore water intake and cooling water systems, sea water desalination plants, and in offshore projects, such as FPSO's, Drillships, and Production Platforms.

This paper will describe the essential design features of the filters, based on the "Bernoulli principle" which result in a fully automated self-cleaning filter system, and choice of materials in a sea water service with high corrosion rates.

A case study will be presented in which stringent project requirements such as substantial flow rates, adaptibility to space limitations and pipe connections had to be observed. Reliability of the system, extended run times without the need for frequent replacement of filter elements, as well as ease of monitoring and operator involvement have been achieved.

352653 Effects of Support on Sulfur Tolerance and Regeneration of Pt Catalysts Measured By Ethylene Hydrogenation and EXAFS

Abstract:

We investigated the effect of support on sulfur tolerance and regenerability under reducing environments by measuring the rate of ethylene hydrogenation, combined with hydrogen chemisorption, and extended x-ray absorption fine structure (EXAFS). Alumina and titania (P25) supports were used to prepare Pt catalysts at 1 wt.%. The catalysts were tested after treatment in H_2S/H_2 at 250 °C followed by regeneration treatments in H_2 at 250 °C, 350 °C and 450 °C. The sulfided catalysts exhibited a 20-27 times decrease in the rate of ethylene hydrogenation and a 4-6 times drop in the Pt surface area. Regenerations up to 450 °C were unable to remove all the sulfur, as shown by the presence of Pt-S bonds by EXAFS. However, a partial recovery of the hydrogenation rate per mole of Pt was observed on sulfided Pt/Al_2O_3 after reduction at 450 °C, while the induction of strong metal support interactions (SMSI) at reduction temperature above 350 °C was observed on Pt/P25, regardless of the presence of sulfur. For Pt/P25, the reversal of the SMSI state together with sulfur removal by mild oxidation suggests that sequential reduction/oxidation treatments may be more effective in restoring the S-free state of TiO_2-supported catalysts.

352799 Improved Operational Efficiency and Reliability through Insulation Materials Selection

Abstract:

This paper will explore the implications of insulation materials selection on the energy efficiency and reliability for Ethylene production units. Considerations will include insulation properties, structural support, installation methods, installation and maintenance costs.

BOILER MACT COMPLIANCE: MERCURY IN FUEL GAS

PATRICK LAINE, WES YONGE, & VIC VICKERY
PEI, HOUSTON, TX 77043

Patrick Laine, Ph.D.
Chief Scientist
Houston, TX
850-866-6686
plaine@pei-tx.com

Wes Yonge, PE
Chief Engineer
Houston, TX
713-806-8396
wyonge@pei-tx.com

James 'Vic' Vickery
Chief Technology Officer
Houston, TX
281-536-0899
vvickery@pei-tx.com

1 Introduction & Objectives

For most ethylene producers the primary objective of this sampling and analysis effort will be to verify and "establish a mercury fuel specification through which gas-fired boilers that use a fuel other than natural gas or refinery gas may be considered Gas 1 unit" as identified in 40 CFR Part 63.7521.(f).

Measurement and Monitoring Requirements: Decision Rules

- ❑If Hg concentration ≤ 50% of Gas 1 specification* then no further testing required
- ❑If Hg concentration > 50% and ≤ 75% of the specification* then semi-annual sampling required
- ❑If Hg concentration > 75% of the specification* then monthly sampling required

*Specification is **40 μg/Sm³**

The goal will be to execute a sampling and analysis campaign that will produce compliance quality data. The key to producing and validating compliance quality data is a well-designed and executed QA/QC protocol. At a minimum, a performance based QA/QC protocol is recommended that includes the following components:

- Sorbent Trap Breakthrough Bed:
  - ❑ Sorbent traps consisting of a minimum of two sorbent beds in series so that the trailing bed can be analyzed to verify that no breakthrough occurred from the leading bed.
- Duplicate Samples:
  - ❑ The sampling system should be configured with two trains to accommodate the collection of simultaneous duplicate sorbent trap samples, in parallel, providing a means of verifying precision in the sample collection process.
- NIST (National Institute of Standards and Technology) Traceable Spike:
  - ❑ Sorbent traps with a third bed (downstream) that is spiked with a NIST traceable spiked mass of mercury serves as a matrix spike and provides a means of verifying that components in the matrix (the fuel gas) are not affecting the trapping efficiency of the sorbent trap.

2 Methods

Sorbent Trap Mercury Sampling Methods Comparison

Method	Description	Method Type: S = Sampling A = Analysis	Evaluation	Sample Collection Phase QA/QC
ISO 6978	Hg in Natural Gas by Thermal Desorption (CVAFS)	S & A	Method calls for low sample volumes can lead to non representative results.	No
EPA Method 30B	Determination of Gaseous Hg Emissions from Coal Fired Combustion Sources using Carbon Sorbent Traps	S	Requires modification (Modified M30B)	Yes
ASTM D6350	Hg S&A in Natural Gas (AFS)	S & A	Method calls for short term samples (i.e., low sample volume) which leads to poor accuracy.	No
ASTM D5954	Hg S&A in Natural Gas (AAS)	S & A	AAS is less sensitive than AFS.	No
EPA Method 1631	Measurement of Hg in Water	A	No	Yes

3 EPA Method 30B QA/QC Performance Criteria

QA/QC Test or Specification	Acceptance Criteria	Frequency
Analyzer calibration	Correlation Coefficient: 0.999	Daily
Transfer efficiency of analyzer	%Transfer Efficiency: 90%–110%	Once per sorbent trap
Initial calibration blank	< 0.5 ng	Once before each calibration
Sorbent Trap Section B breakthrough	≤ 10% of Section A Hg mass, for Hg concentrations > 1μg/dscm; ≤ 20% of Section A Hg mass, for Hg concentrations ≤ 1μg/dscm	At each sample point
Spike Recovery Test	Average recovery between 75% and 125 % for Hg_0	Once per sample point
Paired Sorbent Traps Agreement	≤ 10% RD for Hg concentration > 1μg/dscm; ≤ 20% RD or 0.2 μg/dscm absolute difference for Hg concentrations ≤ 1μg/dscm	At each sample point

4 Equipment

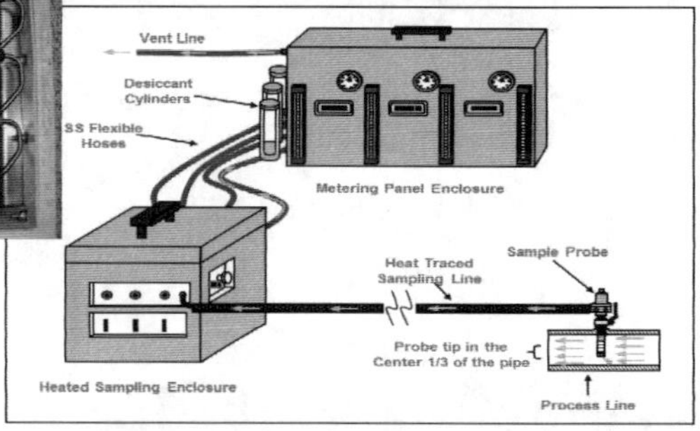

Flow

	Primary Bed	Breakthrough Bed	Spiked Bed

QA/QC Acceptance Criteria:	99% of Total Collected Hg	≤ 10% of Primary Mass	Recovery between 75% and 125%

Gold Trap:

5 Conclusions/PEI Recomendations

- ❑ There are five methods accepted by the Boiler MACT rule. PEI has extensive experience and expertise employing **all five** of these methods.
- ❑ Operators can achieve compliance without submitting a Fuel Analysis Plan by using ASTM D6350, however, we have noticed variable results with this method.
- ❑ The **PEI recommendation** for obtaining the most accurate compliance quality data is to use Modified EPA Method 30B in conjunction with EPA 1631 where traps should be analyzed in a certified ISO 17025 laboratory. The recommended pathway includes a rigorous, performance based QA/QC protocol that validates accuracy and precision of all sample collection procedures.

PEI

351376 Young Professional Simulation Tutorial

Abstract:

18C00 Young Professional Simulation Tutorial

Authors: Naomi Hua and Mike Donahue, Invensys

Process simulation is a complex skill to master. Even with years of experience, a typical user is constantly facing new problems that need to be addressed. There are many issues that can lead to unexpected and unnecessary errors with the use of process simulation tools. Regardless of what simulation program you are using, these issues are common from one program to another. In this presentation, targeted for the young professional, we will discuss the "Most Common Pitfalls in Process Simulation". For each topic, we will discuss various techniques and methods these troublesome problems and prevent their occurrence.

Topics to be discussed include:

1. Erroneous data entry

2. Simulation defaults

3. Bad plant data

4. Thermodynamic method selection

5. Simulation tolerances and convergence

6. Recycle identification and handling

7. Model complexity

8. Difficult or impossible specifications

9. Assay pseudo-component development

10. Assay property blending

1. Erroneous data entry is the most common mistake made when building a process simulation model. This presentation will cover methods that can help ensure proper data entry in your process simulation model, such as checking the data you entered, particularly the units of measure, using "alternate forms" of data input such as copy/paste and data import, and reviewing all warning and error messages.

2. Defaults are methods, values, and conditions that are automatically selected for you by the program when you begin a new simulation model. These defaults will affect the results calculated by your model and every simulation program has them. Most defaults within simulation programs fall under categories such as convergence criteria, thermodynamic methods, tolerance and units of measure. The problem with defaults is unintended results from not fully understanding the effects a particular default has on the model. Take the time to understand the defaults of the program and how these defaults will affect the results. If you do not agree with the defaults selected by the program, most simulation programs will allow the user to change the default.

3. Plant data changes over time. The feed information and process conditions of the unit operations may be different. Not only does plant data change over time, but measurement information is often very inaccurate. This could be due to faulty or poorly calibrated measurement equipment, leaks in lines, sensor failures, and other issues. Because of these changes and inaccuracies over time, you always have to validate the information that is entered in your model to ensure that you are obtaining the proper results. To do so, you can utilize programs that can help with data validation and reconciliation like SimSci's ROMeo online optimization program. With it, you can connect your model to your data historian to constantly import current data measurements into your model and then reconcile these measurements to determine when problems arise within your plant and where those problems are occurring. You will also need to use your engineering judgment to determine if the data is valid. More details will be provided during the presentation.

4. The proper selection of thermodynamic methods for your simulation is the most important decision that you make when modeling a process. The method you select could have a drastic effect upon your results. Often, any number of thermodynamic methods is suitable for a given application. It is up to the user to determine the best method to use based upon knowledge, personal experience, and best judgment. Some important factors that need to be considered are the components being used and the operating conditions.

Oftentimes, the selected thermodynamic method still needs to have additional inputs such as binary interaction parameters to simulate the process accurately.

5. Many of the calculations in process simulation are iterative. When these calculations iterate, the program compares the results calculated from one iteration to the results calculated in the next iteration in order to determine

whether it is converged to a solution or needs to continue calculating. This method of comparing results is where tolerances come into play and simulation programs use tolerances to determine when a solution is reached. One of the major mistakes when modeling a process is not understanding or using the correct tolerance values. A looser tolerance may make the file easier and quicker to converge, but it also reduces the accuracy of the model. A tighter tolerance may be more accurate, but might make it impossible for the model to solve or increase the convergence time exponentially. A common practice is that when you have reached a converged solution with your model, tighten the tolerances within the model and solve it again. If you notice that your answers have changed significantly, you need to continue tightening the tolerance values until your results change little from one iteration to another. You will then know that you have reached an accurate solution for your model.

6. Recycles in process simulation occur when you return to a previously calculated unit operation to recalculate the unit due to a change made elsewhere in the model. They are very common and can often be difficult to converge. Some suggestions on troubleshooting recycle convergence include initially solving your model without the recycles to see how it behaves, providing initial estimates for recycle streams, using tighter tolerances on unit operations within recycle loops than the total recycle tolerances themselves, and ensuring that all components have a way in and out of a recycle by performing mass balances and using purge or makeup streams where necessary.

7. The general rule of thumb when modeling is to start simple and add complexity as you go. Each time you add a few streams or unit operations converge and save the model before moving on. Make sure you justify any additions that you make to the model. You can most likely model your system without including every unit from your PFD in the simulation. When modeling, think about the goal of your simulation and not about literally translating every real-world piece of equipment into the simulation model. Make the judgment call on what is most important and what the purpose is of every unit you add to your simulation. For example, if your real-life process shows a heat exchanger, a pump and a splitter, but all that is necessary is that the temperature, pressure, and separation are achieved properly, consider using a flash drum to accomplish all three.

8. Specifications within process simulation are everywhere and every stream and unit operation you place in your model contains specifications. In its simplest form, a specification is a constraint that you place on the model that the model needs to meet in order to successfully converge. For streams, these are usually temperatures, pressures, flowrates and compositions. For unit operations, they range from process conditions, like temperatures and pressures, to product conditions, like an overhead composition in a column. There are some pitfalls, however, that you can run into with specifications such as azeotropes in columns, high purity specifications, and the possibility of multiple solutions.

9. Petroleum assays are sets of distillation data that are used to characterize all of the components that reside in a given crude sample. When you enter distillation data into a process simulator, the program converts it into components that it can use when modeling your process. The components that are generated are called pseudo-components, and each pseudo-component represents all of the unknown components that would be within a given boiling point range in the crude. Some of the things you will want to ensure when properly developing your assay components are did you use the correct distillation curve basis, did you enter the data correctly, did you use enough cuts and are you using the correct characterization methods .

10. When you have more than 1 assay in a simulation model, you need to be concerned with assay blending and how it will affect the results of your file. Assay blending is the mixing of 2 or more groups of components (i.e. ¨C sets of cuts/pseudo-components) in order to calculate averaged component properties for cuts that are in the same boiling point range as one another. Once these new averaged component properties are determined, the program will use these averages from this point on in the simulation model any time it needs to access component

properties for a given pseudo-component. All component properties are blended, including normal boiling points, specific gravity, molecular weight and refinery inspection properties, and usually the blended component properties are calculated using a weighted average

These topics are based on experience and feedback from end-users to technical support representatives for process simulators. They often represent the most commonly asked questions or areas that most users find to be problematic. While the specific examples highlighted in each topic come from SimSci software, the intent of this tutorial presentation is to be broad-based and is particularly applicable to young professionals seeking to develop their experience with simulation.

Polymeric Sulfonic Acid Composites for Fuel Cell Applications: Effects of Casting Solvent and Sonication Time on Membrane Performance

Jimoh K. Adewole[1], *Abdullah S. Sultan*[1, *], *Amir Al-Ahmed*[2], *and S. M. Javid Zaidi*[3]

[1]*Center for Petroleum & Minerals, King Fahd University of Petroleum & Minerals, Dhahran 31261, Saudi Arabia*

[2]*Center of Research Excellence in Renewable Energy, King Fahd University of Petroleum & Minerals, Dhahran 31261, Saudi Arabia*

[3]*School of Chemical Engineering, The University of Queensland, Australia*

Abstract

Polymer blends comprising sulfonated poly (ether ether ketone) (SPEEK) and 3wt% (2-acrylamido-2-methyl-1-propanesulfonic acid) (PAMPS) were prepared by solution casting method. Two solvents (N,N-Dimethylacetamide (DMAc) and N-Methyl-2-pyrrolidone (NMP)) were uesd separately. Residual solvents in the membrane samples after drying were quantified using thermo gravimetric analysis. Effects of periods of sonication (0 to 90 minutes) and types of solvent on the membrane performance were evaluated interms of morphology and water uptake. An attempt was made to correlate membrane performance with the time of sonication in order to select the most suitable solvent and membrane casting conditions for subsequent membrane preparations. For both solvents, the water uptake was observed to first decrease at 10 minutes of sonication and then increased as the sonication time was increased. The minimum water uptake recorded for DMAc samples was 19.23% at 20 minutes sonication time while that of NMP sample was 24.24% at 10 minutes. However membrane morphology was found to deteriorate with increase in sonication time. No correlation was obtained between the membrane properties and the sonication time. In general, membrane performance was better for membranes samples using DMAc than NMP solvent. Therefore, DMAc was selected as the casting solvent for membrane preparation on which further studies will be carried out.

Keywords: morphology, water uptake, residual solvent, thermogravimetric analysis

Introduction

Water soluble polymers have received academic and industrial interests due to their potential applications in fuel cells and other electrochemical devices (Ye et al. 2012). One of the water soluble polymers that have attracted publication interest is poly (2-acrylamido-2-methyl-1-propanesulfonic acid) (PAMPS) (Florjańczyk et al. 2003). Generally, some of the major performance criteria for fuel cell materials include proton conductivity, water and methanol uptake, good mechanical properties, chemical stability, thermal stability and methanol permeability (Gil et al. 2004; Swier et al. 2005; Ye et al. 2012). PAMPS was reported to posses higher proton conductivity than partially hydrated Nafion. In addition, it has been found to be more tolerant to water fluctuation in high temperature operations. Some of the major drawbacks in the use of PAMPS are its poor mechanical properties, thermal stability, water and methanol uptake and solubility in many solvents (Florjańczyk et al. 2003). One of the approaches that have been proposed to solve these problems is development of polymer composites.

This research paper investigates the synergistic effect of combining water soluble PAMPS with sulfonated poly (ether ether ketone) (SPEEK) which is non - water soluble polymer to prepare composite membranes for application in fuel cells. Specifically, preliminary results on the performnance of the composite were presented. Performance were evaluated based on the composite morphology and water uptake. Effects of sonication time and types of casting solvent on membrane properties were also evaluated.

Material and Method

Material

Sulfonated poly (ether ether ketone) (SPEEK) purcahsed from FuMA-Tech, Germany, and poly (2-acrylamido-2-methyl-1-propanesulfonic acid) (PAMPS) purchased from Aldrich Chemical Co., are the two major polymers used in this research. The SPEEK has a DS of 0.533 and solubility of 99% in DMAc. SPEEK is derived by sulfonation of Poly (ether ether ketone) (PEEK). PEEK is a commercially successful polymer due to its excellent thermal stability and solvent resistance (Mohr et al. 1991). For sometimes, further application of PEEK have been stalled due this solubility issues. By sulfonation, PEEK can be made soluble in most commercial solvents. Importantly, the solubility of SPEEK also depends on its degree of sulfonation (DS). Since, this work involves the use of two polymers, a solvent which will disolve both the PAMPS and SPEEK must be selected. Literature review showed that PAMPS dissolves in N,N-Dimethylformamide (DMF), N,N-Dimethylacetamide (DMAc), propylene carbonate (PC), and water (Florjańczyk et al. 2003). Initial solubility studies that was carried out in our laboratory was done for N,N-Dimethylacetamide (DMAc) and N-Methyl-2-pyrrolidone (NMP). The choice of NMP was due its unique solvent properties and its structtural distiction from DMF and DMAc (Figure 1). Moreover, no literature was found where NMP was used as casting solvent for PAMPS. Thus, DMAc and NMP were used as solvents in this work. All materials were used as received from the manufacturers.

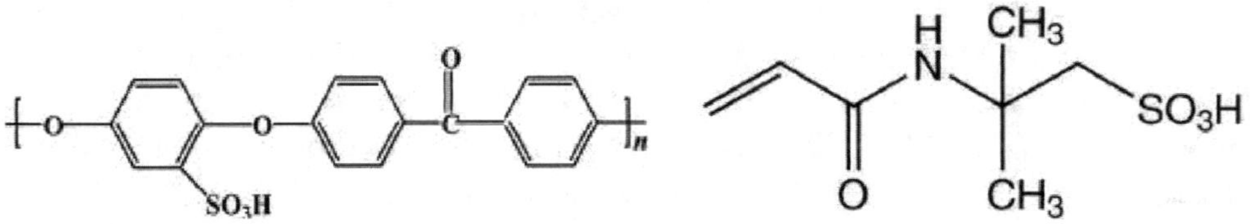

Figure 1 Structure of SPEEK and PAMPS

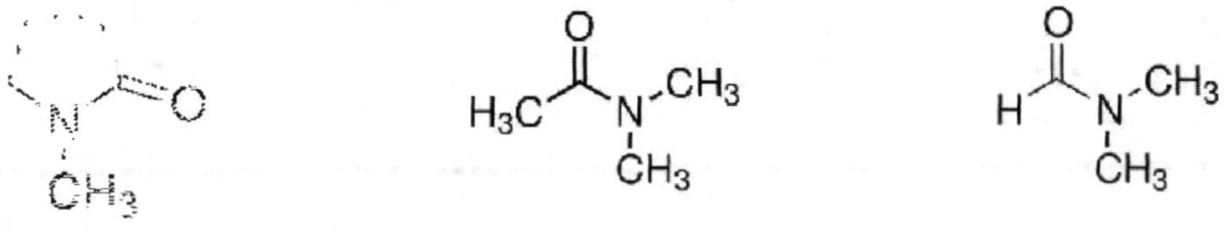

N-Methyl-2-pyrrolidone N,N-Dimethylacetamide N,N-Dimethylformamide

Figure 2 Structure of Solvents

Experimental Procedure

Membrane Preparation

SPEEK was first dissolved in solvents (DMAc and NMP) and stirred for 2hours at 70°C. PAMPS was then added and the blend was stirred for 24hours. The solution was sonicated for various periods of time ranging from 10 to 90minutes using an Ultrasonic bath. In untralsonic bath, the final temperature achieved for a given time of sonication depends on the tempearutre of the laboratory (summer or winter). It is therefoe genrally advisable that maximum temperature of the bath is determined and maintains (as the so - called equilibrium tempearture), when operating continuously under ambient conditions (Santos et al. 2009). During the sonication, equilibrium temperature was maitained by simply filling the bath with water that was preheated to sample temperature.

Membrane Characterization

Water uptake and morphological tests were performed. Water uptake was carried out by first vacuum drying all membrane samples at 100°C until constant weights were obtained. Dried samples were then immersed in deionized water for 24h at room temperature (25°C). Samples were taken out of the water after 24h, wipped to remove surface water and then weighed immediately. Water uptake was calculated using Eq. 1.

$$water - uptake = \frac{W_{wet} - W_{dry}}{W_{dry}} \qquad \text{Eq. 1}$$

Surface morphology of the membrane was done using scanning electron microscopy (SEM). Thermogravimetric analysis (TGA) was done from room temperature to 350°C using TA Q600 analyzer at heating rate of 20°C/min under nitrogen atmosphere with flow rate of 20mL/min using ~ 8 -10mg membranes samples.

Results and Discussion

Membrane Morphology by Scanning Electron Microspcopy (SEM)

Results of surface morphological analysis performed on membrane were shown in Figure 3 and 4. The overall pictures of this figure depicts a dense membrane with some surface defects observed for most of the sonicated samples. For example, Figure 3A showed some kind of layered arrangement with no visible pores at the surface. Also, the PAMPS seems to be well distributed throughout the SPEEK. The micrograph in Figure 3B can be observed to be rough and possesses some uneven ramdom distribution of PAMPS agglomerates indicating that the distribution of PAMPS have been disturbed by the 10minutes sonication. An increase in sonication time to 20 minutes was observed to improve the distribution of PAMPS. This was however accompanied by some cracks believed to be due to the slight increase in the viscousity of the blend solution. The vicousity increase was brought about by the excape of some fraction of the solvent during sonication. As the sonication time was increased, PAMPS was observed to become more evenly distributed and the cracks become more visible and larger in size (Figure 3D to F). SEM images of samples at sonication time of 10 and 30 minutes were reproduced in higher magnification of 10μm (Figure 4) in order to further revealed the effect of sonication on the morphology of the somposites. Similar morphological behaviour has been reported in literature, though for inorganic/organic clusters of nanoclay/epoxy composites membranes (Lam et al. 2005).

Water Uptake

Water uptake have been reported to be affected by the degree of sulfonation, membrane pretreatment, hydration, ambient relative humidity and water temperature (Xing et al. 2004; Wu et al. 2006). Figure 5 showed the effect of sonication time and solvent type on water uptake that was measured at room temperature (25°C). For both solvents, water uptake is same at zero sonication time. There was however a further decrease in the water uptake for NMP at 20 minutes sonication time before it beging to increase. Thus, the best sonication time for both solvent lied between 10 and 20 minutes with respect to water uptake. Interestingly, this result correspond to that reported by Lam (2005) on the effect of ultrasonication on microhardness of composites.

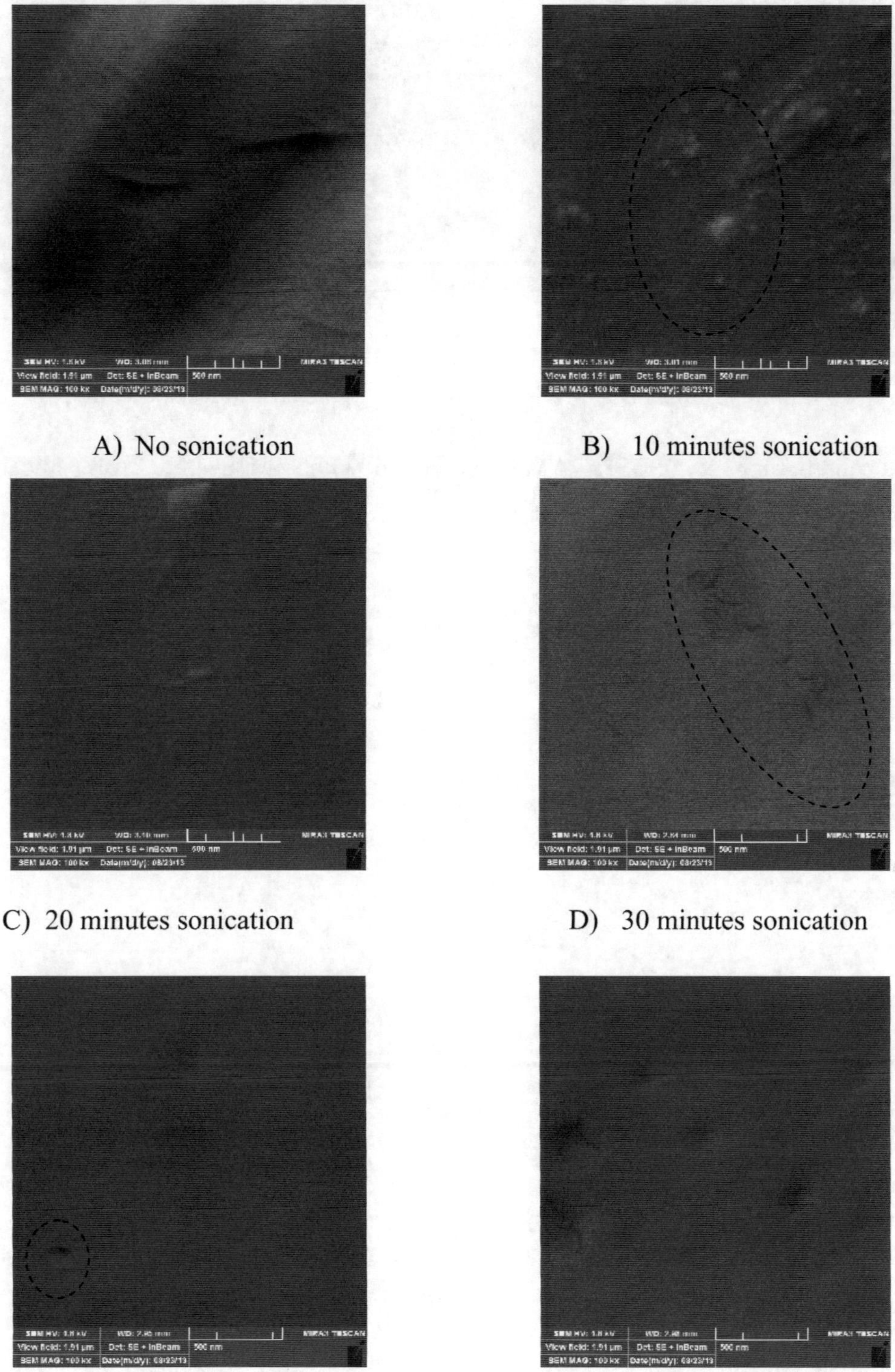

A) No sonication

B) 10 minutes sonication

C) 20 minutes sonication

D) 30 minutes sonication

E) 60 minutes sonication

F) 90 minutes sonication

Figure 3 SEM micrograph of SPEEK-PAMS composite with various times of sonication with magnification of 500nm

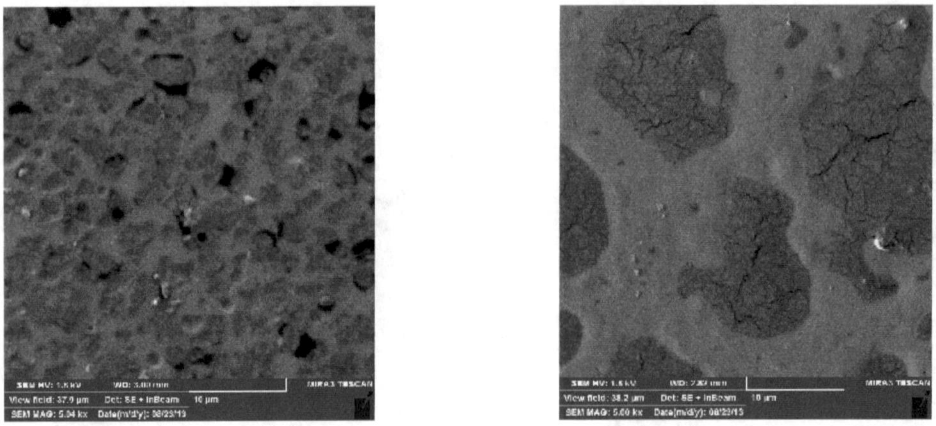

A) 10 minutes sonication B) 30 minutes sonication

Figure 4 SEM micrograph of SPEEK-PAMS composite with various times of sonication with magnification of 10μm

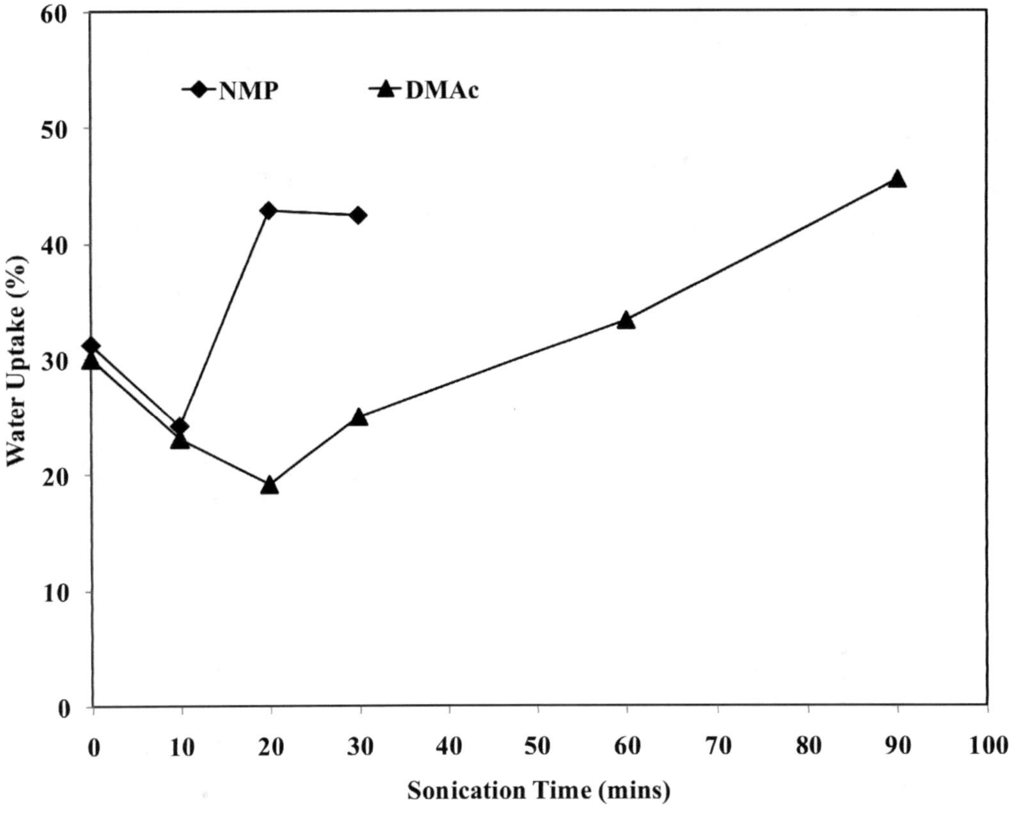

Figure 5 Water uptake of SPEEK-PAMS (3wt%) composite with variuos sonication times

Thermogravimetric analysis (TGA)

The difference in membrane performance resulting from the use different casting solvents have been attributed to the presence of residual solvents in the membrane samples after evaporation (Joly et al.

1999). TGA was thus used to quantify the residual solvent and evaluate its effect on the membrane performance as has been previously demonstrated in lieterature (Recio et al. 2007; Iqbal et al. 2008). Figure 6 shows the TGA curve for samples prepared from NMP and DMAc with different post vacuum heat treatment conditions. From this figure, weight loss can be attributed to evaporation of variety of substances, including solvents, from the samples (Rich et al. 1997). Table 2 contains the solvent and its boiling points, and weigth loss at the boiling point range that were extracted from TGA results. For sample heat treated for 24hrs, the overall percentage weight loss up to 204°C was about 8wt% of membrane sample for NMP while it is about 6.3wt% for DMAc. These will definitely include compnenents such as water, solvents and other unidentified substances. To quantify residual solvents, weight loss was estimated between 192 -214°C for SPEEK-PAMPS-NMP and between 154 and 176 for SPEEK-PAMPS-DMAc. The results obtained showed that the amount of the residual solvent in NMP is larger than that of DMAc by 1.25wt% (Table 1). To further investigate if this difference is responsible for higher value of water uptake for the NMP sample, membrane samples made from both solvents were further heat treated under vacuum for 72hrs. Residual solvent after this treatment was slightly higher in DMAc than NMP. Despite this, the value of water uptake of the NMP membrane is still higher than that of DMAc by approximately the same value as before the heat treatment (1.71wt%).

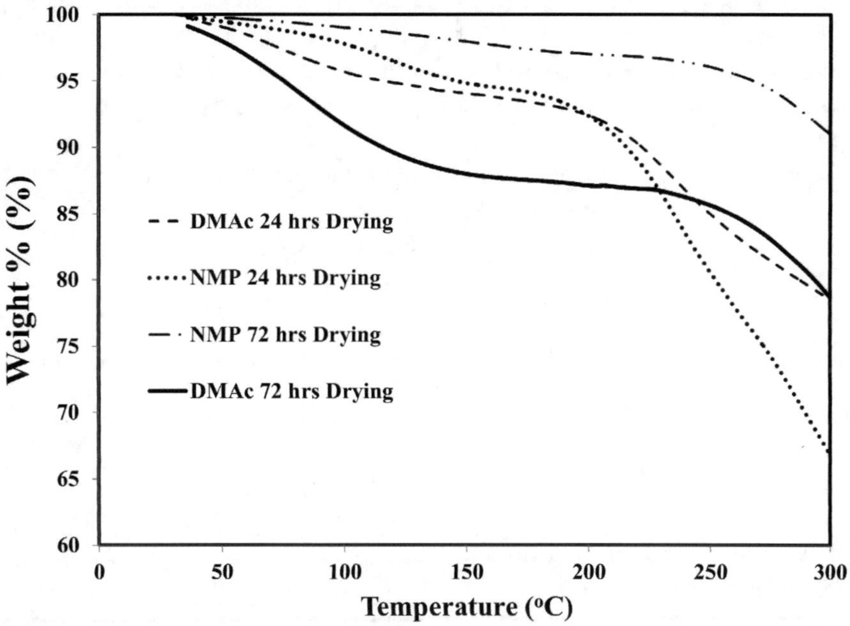

Figure 6 TGA of SPEEK-PAMPS-NMP and SPEEK-PAMPS-DMAc

Table 1 Residual solvent in SPEEK-PAMPS composites membrane samples

Membrane Sample	Solvent	B. Pt. (°C)*	Residual Solvent (wt%)	Water Uptake
DMAc 24 hrs Drying	DMAc	164.5 – 166	0.57	30.00
NMP 24 hrs Drying	NMP	202 – 204	2.85	31.25
DMAc 72 hrs Drying	DMAc	164.5 – 166	0.367	21.21
NMP 72 hrs Drying	NMP	202 – 204	0.217	22.92

*Boiling Points of Solvents

On a general note, results from Figire 1 – 5 showed that the best perfomance can be obtained for variety of sonication time depending the property of preference. Also, performance of DMAc solvent was observed to be better than that of NMP on the basis of water uptake. For this reason, DMAc was selected as the casting solvent for membrane preparation on which further studies will be carried out. Water uptake of DMAc membrane, prepared without sonication, was comapred with those reported in previuos literature. Table 2 illustrates the comparison between SPEEK-PAMPS and other relevant membranes previuosly reported in literature. In addition, this table serve to provide insight into the possible potentials of the composites. From this table, it was observed that the water uptake for SPEEK-PAMPS is within the range of values of other membranes on the table. Its water uptake is less than that of Nafion 112 which is a commercial membrane. The differences observed between the water uptake values of SPEEK – PAMPS and other membranes from the table can be explained based on three important factors - SO_3H group, the degree of sulfonation and experimental temperature (Gil et al. 2004; Shan et al. 2006; Wu et al. 2006). Water uptake of SPEEK – PAMPS is higher due to the presence of SO_3H group in the PAMPS molecules (Figure 2). SO_3H group have been reported to enhance membrane acidity and hydrophilic properties (Shan et al. 2006). Infact, the higher the SO_3H group the larger the water uptake. The degree of sulfonation has been reported to affect the water uptake of membranes. The higher the degree of sulfonation, the larger the value of water uptake (Gil et al. 2004). Thus, water uptake of SPEEK – PAMPS is lower than SPEEK -3 due their different degree of sulfonation, 53.3 and 72% respectively and the experimental temperature 24 and 60°C, respectively. The experimental temperature of SPEEK – 2 is close to that of the composite. However, the results cannot be compared in this case since the DS of SPEEK – 2 was not known. Above explanation is also true for the values of water uptake for SPEEK – 1.

Table 2 Comparison of water uptake values obtained with those from Literature (Xing et al. 2005)

Membrane	Water Uptake (wt%)	Experiment Temperature (°C)	**D S (%)	Reference
Nafion 112	22.85	N/A	-	(Xing et al. 2005)
SPEEK – 1*	16.00	30	45	(Ren et al. 2005)
SPEEK – 2*	44.90	25	N/A	(Zhong et al. 2008)
SPEEK – 3*	85.00	60	72	(Wu et al. 2006)
SPEEK– 4*	15.00	N/A	N/A	(Handayani and Dewi 2011)
SPEEK/PAMPS (3wt%)	21.21			This work

*The numbering have been included for explanation purposes.
**DS means Degree of Sulfonation

Conclusion and Recommendations

Blend of SPEEK and PAMPS was prepared by solvent casting method and its performance was evaluated with respect to its morphology and water uptake. Effect of sonication time (10 – 90minutes) and solvent types (DMAc and NMP) were also investigated. Performance of membrane prepared using DMAc was observed to be better than that of NMP. There was no correlation between the sonication time and membrane properties, but it was observed that the sonication deteriorate the membrane morphology. Membrane prepared without sonication shows better surface morphology, whereas the best water uptake was observed between 10 and 20 minutes. Thus, other properties need to be investigated in order decide the best sonication time. Therefore, further investigation such as thermal stability, methanol permeability, ion exchange capacity and gas (O_2 and H_2) permeability of the membrane are recommended for a complete evaluation of the membrane performance for direct methanol fuel cell (DMFC) application. All these investigations are in progress in our laboratory.

Acknowledgement

Authors would like to thank the National Science, Technology and Innovation Program (NSTIP) of KACST for funding this research under project number 10-ENE1374-04. Also, authors extend their appreciation to Center of Petroleum & Minearls, Center of Excellence in Renewable Energy, Mr. Dan Jerry D.Cortes of the Nanotechnology Laboratory, Chemical Engineering Department, King Fahd University of Petroeleum & Minerals, Saudi Arabis for their supports.

REFERENCE

Florjańczyk, Z., et al. (2003). "Proton conducting electrolytes based on poly(2-acrylamido-2-methyl-1-propanesulfonic acid)." Electrochimica Acta **48**(14–16): 2201-2206.

Gil, M., et al. (2004). "Direct synthesis of sulfonated aromatic poly(ether ether ketone) proton exchange membranes for fuel cell applications." Journal of Membrane Science **234**(1–2): 75-81.

Handayani, S. and E. L. Dewi (2011). "Influence of Silica/Sulfonated Polyether-Ether Ketone as Polymer Electrolyte Membrane for Hydrogen Fueled Proton Exchange Membrane Fuel Cells." Internat. J. of Sci. and Eng. **2**(2): 26-30.

Iqbal, M., et al. (2008). "Solvent effect on morphology and CO2/CH4 separation performance of asymmetric polycarbonate membranes." J. Membr. Sci. **318**(1–2): 167-175.

Joly, C., et al. (1999). "Residual solvent effect on the permeation properties of fluorinated polyimide films." Separation and Purification Technology **16**(1): 47-54.

Lam, C.-k., et al. (2005). "Effect of ultrasound sonication in nanoclay clusters of nanoclay/epoxy composites." Materials Letters **59**(11): 1369-1372.

Mohr, J. M., et al. (1991). "Gas transport properties of a series of poly(ether ketone) polymers." Polymer **32, 13**: 2387-2394

Recio, R., et al. (2007). "Gas separation of 6FDA–6FpDA membranes: Effect of the solvent on polymer surfaces and permselectivity." J. Membr. Sci. **293**(1–2): 22-28.

Ren, S., et al. (2005). "Surface modification of sulfonated poly(ether ether ketone) membranes using Nafion solution for direct methanol fuel cells." Journal of Membrane Science **247**(1–2): 59-63.

Rich, D. C., et al. (1997). "Effect of cure conditions on Probimide 32 polyamide-imide." Journal of Applied Polymer Science **63**(9): 1113-1126.

Santos, H. M., et al. (2009). The Power of Ultrasound. Ultrasound in Chemistry: Analytical Applications. J.-L. Capelo-Martinez. Weinheim, Wiley-VCH.

Shan, J., et al. (2006). "Sulfonated polyether ether ketone (PEEK-WC)/phosphotungstic acid composite: Preparation and characterization of the fuel cell membranes." Pure Appl. Chem **78**(9): 1781 - 1791.

Swier, S., et al. (2005). "Sulfonated poly(ether ketone ketone) ionomers as proton exchange membranes." Polymer Engineering & Science **45**(8): 1081-1091.

Wu, H.-L., et al. (2006). "Preparation and characterization of poly(ether sulfone)/sulfonated poly(ether ether ketone) blend membranes." European Polymer Journal **42**(7): 1688-1695.

Xing, D. M., et al. (2005). "Characterization of Sulfonated Poly(Ether Ether Ketone)/Polytetrafluoroethylene Composite Membranes for Fuel Cell Applications." Fuel Cells **5**(3): 406-411.

Xing, P., et al. (2004). "Synthesis and characterization of sulfonated poly(ether ether ketone) for proton exchange membranes." Journal of Membrane Science **229**(1–2): 95-106.

Ye, Y.-S., et al. (2012). "Water Soluble Polymers as Proton Exchange Membranes for Fuel Cells." Polymers **4**(2): 913-963.

Zhong, S., et al. (2008). "Modification of sulfonated poly(ether ether ketone) proton exchange membrane for reducing methanol crossover." Journal of Power Sources **180**(1): 23-28.

Fixed Plant Facility Siting – A Mega Project Lifecycle Framework

M.F. Haider
mech_haider@yahoo.com
mhaider@suncor.com

Prepared for Presentation at
American Institute of Chemical Engineers
2014 Spring Meeting
10th Global Congress on Process Safety
New Orleans, LA
March 30 – April 2, 2014

Fixed Plant Facility Siting – A Mega Project Lifecycle Framework

M.F. Haider
mhaider@suncor.com
mech_haider@yahoo.com

Keywords: Facility Siting, Project Management, Mega Project, Consequence Analysis (CA), Quantitative Risk Assessment (QRA), Adaptive, Change driven, Agile methodology

Abstract

Facility Siting based on tools such as Consequence Analysis (CA) and Quantitative Risk Assessment (QRA) is often times successfully applied for Brownfield operating unit projects, projects with relatively smaller footprints and for projects lasting relatively short time-periods. Challenge remains to apply and realize the full potential and value of Facility Siting for Mega-Projects. This is mostly due to project complexity and length – this paper will introduce a Management Framework to successfully apply CA and QRA as part of Facility Siting for Mega-Projects. The paper will also outline key strategies to ensure successful (i.e. value adding) implementation of facility siting, suggest a list of activities, associated sequence and pitfalls to avoid. Project Management team (e.g. Project Sponsor, Project Engineers, Managers) of both owner/operator and contractor/vendor may find this paper more useful as it provides a broad overview (i.e. when to initiate, how to plan, execute and close such assignment to maximize value) including real life examples of facility siting process over the project lifecycle.

1. Introduction

The driving force for process safety has been primarily based on catastrophic events [1]. With an increasing number of incidents, the process industry and governments started to take measures such as Facility Siting initiatives with hope to minimize loss of life and property. In the 1970s and 1980s, some of the world's most tragic industrial accidents took place. The Flixborough explosion in 1974 was reinforced by that of the Seveso tragedy in 1976 [2]. The Macondo blowout in the Gulf of Mexico on April 20, 2010, killed 11 employees and led to an uncontrolled oil spill lasting 87 days. This blowout was the most significant offshore incident in the US to-date. Finally, the Fukushima Daiichi nuclear plant incident in Japan in March 2011 drew the attention of the global process and power industries and encouraged them to incorporate natural disaster risks in a hazard analysis study [3]. As part of inherently safer design, many companies now-a-days apply CA and QRA as part of facility siting for their new projects. Due to the scarcity of information at the beginning, the studies in many cases start later in the project lifecycle for them to be effective and value adding. This paper will focus more on how to manage such initiatives for new Mega-Projects over the technical intricacies of the CA/QRA processes. The intention of the paper is to enhance the Body of Knowledge publicly available for management team in as much non technical languaga as possible to make their greenfiled Mega-Project initatives a success i.e. safer for people, environment and property.

So, what is different for industrial Mega-Projects?

- Industrial Megaprojects, as the name implies, are very large. To provide a simple and simply applied definition, we are defining a megaproject as any project with a total capital cost of more than $1 billion (U.S. dollars) as measured on January 1, 2003 [4].
- Most often engineering and construction activities run in parallel and not in sequence
- Multiple sub-projects under the 'Mega-Project umbrella' run in parallel
- Complex – design is almost 'never' complete in early stages – goes through many iterations
- Progressive elaboration is evident
- Scarcity of critical workforce with mega-project experience
- Scarcity of information – requires project team to make assumptions for CA/QRA
- Timeline: 2-5 years

2. Key Strategies

2.1 Facility Siting Charter

It is not only important or good to do but critical that the stakeholders need to identify, understand and agree on the purpose and value to conduct facility siting for their projects. The Author's personal experience over the years has been in order to maximize value of Facility Siting the project team has to start 'early' to develop project Charter and start with the end in mind. As can be seen from the following generalized Figure 1 that as the company starts to appraise opportunities, it is

Figure 1: Stages and Milestones to apply CA/QRA in a Mega-Project

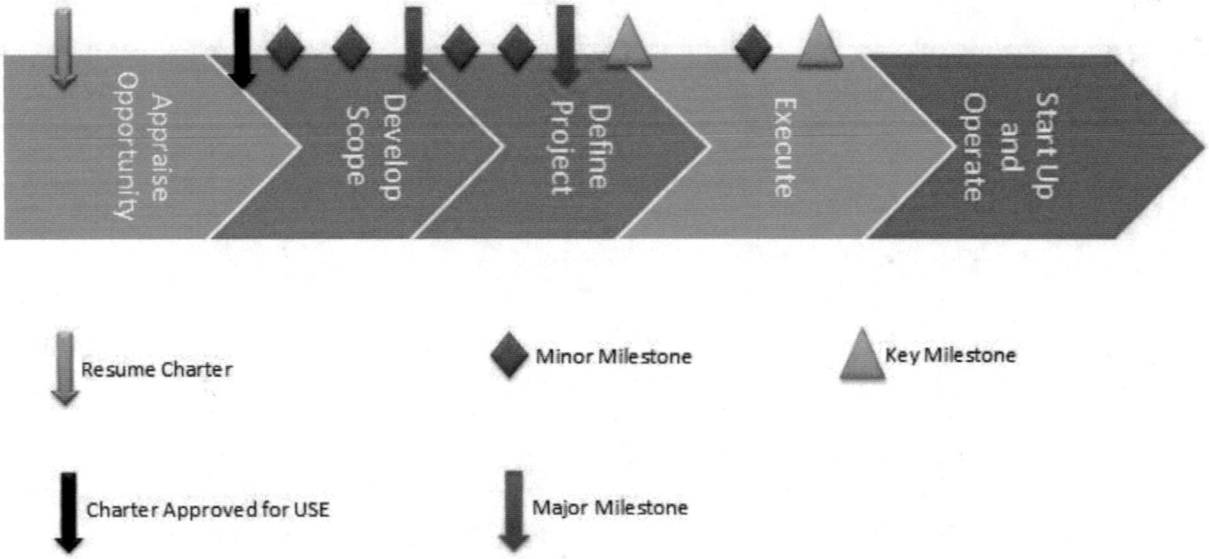

prudent to resume developing the Facility Siting Charter by bringing in the stakeholders together. Figure 1 shows a generalized project lifecycle for a typical Petro-Chemical Mega-Project. The

Figure also shows that the Key Deliverable is developed through a successive series of repeated cycles over the project life-cycle. The Milestones will be defined later in the paper. Part of starting with the end in mind is to clearly outlining ultimate goal/s, e.g. vision is to design a safer facility and utilize Mitigated QRA Report to formally document and communicate that facility is within the Risk Threshold or justification for accepting risk profile as is by Senior Management. The Vision in the Charter also provides opportunity to strike a balance between business goals and climbing the Risk Mountain – at each stage of increasing safety (decreasing risk). As part of striking a balance, the Project Engineer has to Identify options up-front if and/or when it becomes apparent that aggregate Residual Risk will not be within the Risk Threshold or acceptable so that team can put more effort in developing justification for either accepting or rejecting the risk profile by Management

Figure 2: Strike a balance between Business goal and Climbing the *expensive* Risk Mountain

2.2 *Management Strategy*

In order to successfully manage Facility Siting, CA and QRA for a Mega-Project, the stakeholders have to align on a management strategy. There are two primary management approaches for the purpose of managing such work:

1. Predictive or Plan driven
2. Adaptive or Change driven

The Author's past experience has been if the organization undertook very 'similar' projects in the near past (<5 years), critical experienced workforce is retained, Lessons Learned is readily available, and most importantly, the past project is still considered to be a success, it is advisable to use Plan driven approach. On the other hand, if the organization is introducing new technology, has limited past experience with similar project, involves relatively high number of stakeholders, capital cost > $1B, project life spans >2 years, it is recommended to utilize Change driven approach to maximize value. It is to be noted that it is not essential to meet all the suggested criteria mentioned above to pick one management strategy over the other.

2.2.1 Benefits of Adaptive or Change Driven Approach

In Adaptive/Change Driven Planning, project team intentionally repeats one or more project activities as the project team's understanding of the product increases ……….Also known as …agile methods …..intended to respond to high levels of change and ongoing stakeholder involvement [5]. The heart of Facility Siting involves making timely but informed decision. Some of the characteristics of this framework:

1. To add small incremental improvements that will deliver value to stakeholders
2. Milestones develop the product (e.g. Mitigated QRA Report) through a series of repeated cycles, while increments successively add to the functionality of the product
3. For each Milestone, a deliverable (read 'model') or set of deliverables will be completed. Future iterations may enhance those deliverables or create new ones. Each iteration incrementally builds the deliverables until the exit criteria for the phase are met, allowing the project team to incorporate feedback.

The biggest challenge with Mega-Project is scarcity of key information needed as inputs to develop scenarios and models. Some of the critical benefits, a project may gain by applying Change Driven approach:

1. The approach acts as Leading Indicator on Process Safety for Plant Layout - allows for multiple opportunities to make critical decisions on plant layout over the project life-cycle capitalizing on what is known at a given time. This lets the project team to use "what they know", to ensure "what they don't know" falls into the right place
2. Opportunity to make decisions over the project lifecycle not only works to save life but also helps with striking a balance between business needs and climbing the risk mountain
3. Aides design-development with progressive elaboration while producing final CA/QRA deliverable for the project

2.3 Execution Strategy

The main idea of Change Driven process starts with decomposing the known preliminary layout and process information into a set of requirements, work to be performed and their priority sequence, sometimes referred to as a *Product Backlog*.

It is advisable that the owner/operator forms a multi-use agreement (contract) with the chosen vendor. This way the plant areas under the Mega-Project umbrella can get their specific work done while the CA/QRA Project Team provides overall oversight and guidance.

At the beginning of an iteration, the team will work to determine how many of the highest priority items on the backlog list can be delivered within the next iteration. The time-length of each iteration should be less than 4 weeks. As shown in Figure 1, during "Develop Scope" and "Define Project", the length of each iteration could be even less than 2 weeks. This iterations (otherwise known as "Minor Milestone" as shown on Figure 1) will develop scenarios, assumptions and preliminary CA/QRA models. Assumptions are required due to lack of available quality information in the early stages of a project.

At the end of each iteration, the Minor Milestone/deliverable should be ready for review by the customers/stakeholders. This does not mean that the customer is required to accept delivery, just that the product should not include unfinished and/or incomplete features.

The sponsor and customer representatives should be continuously engaged with the other project stakeholders to provide feedback to Contractor on deliverables as they are created and to ensure that the *Product Backlog* reflects Sponsor/stakeholders' current needs. Below, you will find two examples using two different management strategies. For confidentiality purpose, the specific name/location is kept generic and few items were modified to suit.

2.3.1 Project X Facility Siting Using Change Driven Approach

A CA/QRA team was given the assignment to support design development during "Develop Scope" phase (Figure 1). The Sponsors already selected the site for the project. But the fixed plants were yet to be located on the site. The Vision for the initiative was for the QRA result to come under the organization Risk Threshold. Multiple EPC's were given to engineer different plant areas. The CA/QRA team selected a vendor to support them with defining the *Product Backlog* based on the preliminary Process Design Basis, intended product the facility is expected to sell, the feedstock necessary etc. The CA/QRA team had prioritized the Product Backlog based on hazard level of feedstock, preliminary PHA schedule and project schedule. This also involved obtaining feedback from different Process Engineering and Operations teams to collect necessary inputs to put together the assumptions and scenarios to be modeled within the 1st iteration.

Out of 25+ scenarios, only 7 perceived to be the most hazardous were modeled for CA that gave ample input (e.g. explosion, toxic and fire contours) for the CA/QRA team to work with the stakeholders to recommend location for specific process units on the given site. Note that, CA/QRA team integrated the potential consequence to understand local incident and global consequence events. At this time QRA was not performed. This was used as input for minor modification of preliminary plant layout. It is to be noted that the site plan was still preliminary after the first iteration but assisted different EPC accountable for their specific area to recognize where to locate the specific plants i.e. finalized key plans for the overall site plan.

With further design-development, more refined information became available from each EPC. This new information led to 130+ scenarios and was used for further CA/QRA iterations that created either new model and/or embellished the past ones. Each EPC was accountable for facility siting within their boundary while the CA/QRA team provided technical oversight and integration support for the overall project. The Sponsor/stakeholders were engaged on a regular basis to ensure that the Product Backlog reflects future modeling requirements to properly site facilities e.g. Fire Water Pump house, Tank Farms, site Medical Centre etc. Early engagement of CA/QRA team and higher management willingness to incorporate their feedback with the progressive elaboration of the project led to QRA result meeting the organization risk threshold and ultimately safer facility. Changes suggested based on CA/QRA work in earlier stage (i.e. Develop Scope, Define Project) were easier to incorporate into the design.

2.3.2 Project Y Facility Siting Using Plan Driven Approach

The sponsors for project Y selected the site for the project and decided to hold on to not perform CA/QRA work with the progressive elaboration of the project activities. This was mostly decided based on the fact that not all information was available to properly conduct the 'studies'. Fundamentally, the sponsors looked CA/QRA as deliverables requiring a more 'complete' plan. Once the Develop Scope (as referred in Figure 1) is almost complete, the sponsors decided to engage CA/QRA team and a vendor. By this time, the project had significantly high number of stakeholders. Many of the stakeholders were not very familiar with CA/QRA and the value it could bring to them. There was not much time to schedule and train the stakeholders as well to bring them up to speed. As a result, the stakeholders were unable to align on a Vision. The Vendor was asked to provide explosion, toxic and fire contour maps for specific scenarios and locations that were almost 'fixed' on the site plan at that point in time. The resultant contour maps showed significant consequence on nearby facility with over 700+ people. It was also noted that the plant area consuming the most hazardous substance would impact the nearby facilities significantly from consequence perspective. It should be noted that the result was only based on a selected few scenarios. By the time the project reached the "Execution" phase, more scenarios were added as part of the scope. But it was already too late to incorporate most of the recommended changes resulting from CA/QRA work. The project went through several area by area risk reduction workshops only to accept (by the company executives per organization risk matrix) the aggregate residual risk level which was also found to be not within the organization risk threshold.

2.3.3 Importance of Starting Early

Figure 3: Impact of variable based on project time [5]

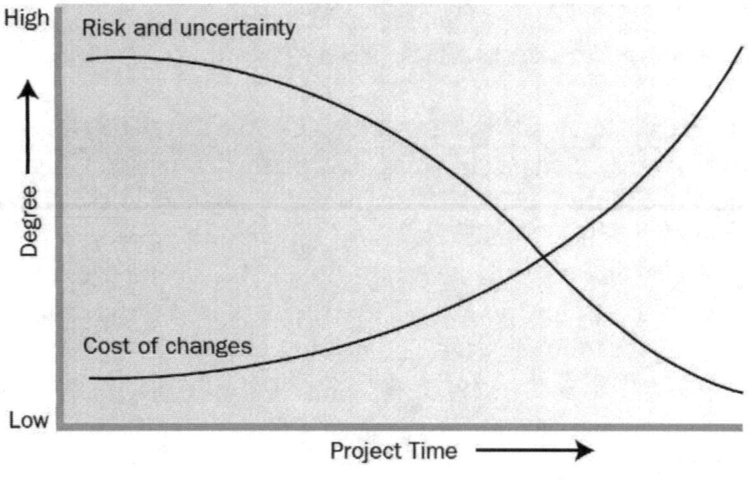

The above two examples highlight the importance of starting CA/QRA work early enough to realize and maximize the value CA/QRA brings to a Mega-Project. The high-level view provides common frame of reference for comparing projects for CA/QRA initiative, even if the projects were dissimilar. Change driven methodology allows for applying the known information at early stage in a way that supports laying out plant facilities to reduce risk and thus maximize potential value. While these characteristics remain present to some extent in almost all project life cycles, they are not always present to the same degree [5]. Change Driven framework, in particular, keeps stakeholder influences higher and the costs of changes lower especially in the early stages of the life cycle than in predictive/plan life cycles. The Figure 3 highlights a well-known fact of project management that even though risk and uncertainty remains high in the early stages of a project, cost of incorporating changes remains much lower. This translates into a key opportunity for project stakeholders to align on laying out the key areas on a site plan intelligently to reduce risk and maximize value by applying tools such as CA and QRA.

2.4 Key Decision Points

Facility Siting, from a management point of view, is all about bringing the stakeholders closer and find the 'right' spot for the facilities. Finding the 'right' spots requires making timely decisions. The challenge has always been to make timely decisions as the stakeholders more often find themselves forced to make those decisions based on information that are frequently of poor quality and at times uncertain, especially in the early stages of a project. At times, due to the scarcity of information, stakeholders rely on their past experience to make those decisions on plant layout. As a result, depending on the individual experience, two different teams for a project area may come up with two very different plant layout. CA/QRA work, if started early enough, will also depend on input that most often requires embellishment through assumptions to create needed scenarios and models. Even with this limitation, CA/QRA provides a key 'input' in allocating real estate for key facilities on a plant site at early stage of a project lifecycle. With each iteration and deliverable 'Change driven' methodology allows for multiple opportunities to make critical decisions on plant layout over the project life-cycle. It allows the stakeholders to decide when to 'sharpen the pencil' (e.g. PHAST to Computational Fluid Dynamics model), relocate facility or accept the consequence and/or risk, 'pull the plug' on further modeling/refinement and start developing justification, if required, to recommend management to accept/reject the aggregate risk profile.

2.5 Alignment with Process Hazard Analysis (PHA) Planning

CA/QRA results are a useful input to PHA (e.g. HAZOP, LOPA) activities. This is particularly true as projects move forward after establishing the plant layout on a high level and starts to conduct "Define Project" (Figure 1). PHA activities may also result in recommendations (i.e. new scenarios) that goes into the *Product Backlog* and requires project team to develop new models or refine old ones. Mega-projects have multiple sub-projects running in parallel most of the time, having an Integrated Schedule for all the reviews is a good strategy so that the right stakeholders are booked for the sessions. The integrated PHA schedule also supports prioritization of *Product Backlog* for CA/QRA activities over the lifecycle of the project.

2.6 Integration Strategy

It is critical for a Mega-Project involving multiple plant areas/silos to understand the BIG picture so that site risk can be dealt with on an aggregate basis. There is strong possibility of one area to impact other plants in the vicinity. It is recommended to review integration aspect and aggregated results on a regular basis. The following Figure 4 shows that the origin of an incident may reside in one particular plant area, while it may impact other areas in the vicinity.

Figure 4: Consequence of an incident in one area impacting multiple areas

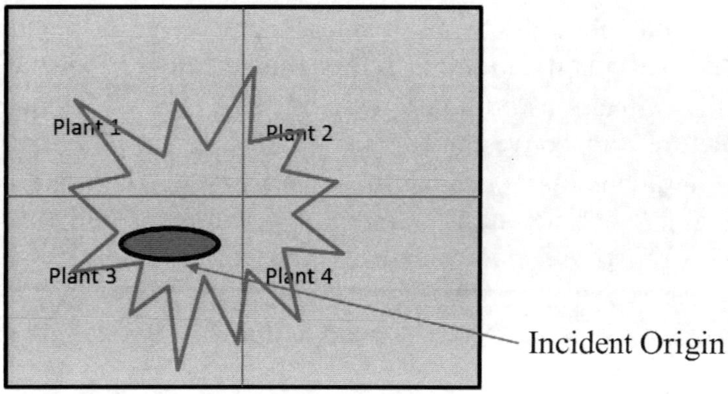

Recommended Strategy:

- Setup Strong Center - Task Force Single Point of Contact (SPOC)
- Collaborate – SPOC will promote Collaboration among Plants/Areas

3. List of Activities (suggested)

3.1 Overall Sequence

The CA/QRA work typically should start with the approved Charter as shown on Figure 1. A project Task Force i.e. SPOC is formed comprising of Project Engineer, Operations rep, Process rep and CA/QRA Technical Expert. The SPOC is ultimately accountable to produce the CA/QRA deliverable studies but also acts to support the design development and site plan finalization efforts in the early stages of the project.

The team collects 'input' such as Preliminary site Layout and design information, works with area process and operations team to identify scenarios, develop assumptions for known-unknowns and unknown-unknowns, develop modeling 'Product backlog', prioritize backlog based on PHA schedule and produce modeling output (i.e. deliverable) for stakeholder use. The stakeholder usually has the option of accepting or rejecting consequence. If the result of an iteration is acceptable within area causing the incident – communicate to outside stakeholders to ensure there is no cross boundary impact. If the result of the iteration is NOT acceptable to outside of the area – use the result as input to PHA for the areas to mitigate. Collaboration recommended between origin area and impacted area/s to identify appropriate mitigations. There is also a possibility that the consequence resulting from the iteration is not acceptable to the area causing the incident, in that case the area will initiate a PHA to mitigate the consequence. As part of developing mitigations, consider the following:

- Develop QRA if appropriate
- Perform Building Assessments for Critical Buildings
- Apply Cost-Benefit analysis to optimize risks/opportunities

- Determine Safe Haven locations and their characteristics

If Mitigation is not possible from Business perspective – perform As Low As Reasonably Practicable (ALARP) justification for Senior Management to formally accept the posed risk

- To apply mitigations on a site wide basis, review Milestone deliverables by taking on a BIG picture perspective
- Future iterations may enhance those deliverables or create new ones as new information is available. Each iteration incrementally builds the deliverables to produce the Mitigated CA/QRA report to support Facility Siting for the stage

The overall effort is expected to produce a maximum of two single reports for the overall site:

1. Mitigated version of QRA Report
2. and Consolidated CA Report

It is recommended to develop as-built, pre-start up CA/QRA reports prior to start-up. Table 1 and 2 outlines typical activities performed by SPOC and plant areas respectively once key process areas are established/finalized on Site Plan. Please note that the activities are project dependent and may warrant modifications to suit specific situation.

Table 1. SPOC Sequence of Activities to conduct Facility Siting

Initiative	Key Steps
Setup Facility Siting Team	Sponsor to support to setup a team (i.e. Project Engineer, Technical Expert, Operation, Process representative) to act as SPOC
Integration	Provide functional (technical and management) guidance to Areas
	Assess cross boundary impact/opportunity
	Communicate incremental outputs after each Area PHA/Risk Reduction
IFA versions (prior to "Define Project" stage)	Consolidated CA Report
	Mitigated and Integrated QRA Report
IFD CA/QRA (prior to "Execution" stage)	Re-validate IFA CA/QRA post PHA and vendor PHA
As-built versions	Post-oil version

3.2 *Typical Area Activities*

Table 2. Area (Typical) Sequence of Activities to conduct Facility Siting

Initiative	Key Steps
Plant 1 (Typical)	Perform Modeling and Calculations
	Conduct Risk Reduction Workshop (if appropriate) – Apply mitigations
	Incremental output for use based on reduced risk level
	Assess Buildings/facilities
	Identify Safe Haven/shelter Locations (if required)
	Estimate Area Project Scope impact
	Perform area scope updates/changes e.g. passive fire protection requirement
	Carry out QRA (if appropriate)
	Update Product Backlog
	Develop/update contract to support area specific activities

4. Pitfalls to Avoid

The following observations were noted by the Author during his time to manage Facility Siting scope:

1. Facility Siting does not replace good and common sense engineering practice e.g. locating a plant that uses high volume of hazardous material will significantly increase risks to other plants/areas (e.g. camp) within the vicinity
2. Evaluate effectiveness of 'Cost Reduction' philosophies as opposed to added risk e.g. different modularization techniques may decrease the plant footprint by reducing pipe lengths but makes the plant more compact and likely to increase severity of explosion consequence
3. Avoid developing models with hope to realize risk reduction
4. Sharpening the pencil for models most likely may not reduce risk
5. Avoid jumping into CFD or other similar methods without evaluating cost-benefit for undertaking such effort
6. Utilization of QRA results is much more controversial than the methodology. Watson [6] has suggested that QRA should be considered as an argument, rather than a declaration of truth. In this view – focus on accuracy over precision
7. CA/QRA Task Force should be part of Management of Change (MOC) process
8. Evaluate carefully before locating facilities e.g. tank farms close to main entrance/egress as part of Facility Siting

9. Keep the focus on VISION for the overall effort to avoid scope creep
10. Note the difference between Facility Siting as a Deliverable and tool for design-development activities

5. Conclusion

This paper presented Adaptive/Change driven framework to manage initiatives like facility siting/CA/QRA as part of Mega-Projects. Change driven framework does not mean "lack of planning or no planning". As demonstrated in the paper, this methodology starts with a vision and then develops the ultimate product both iteratively and incrementally when it is possible to define small incremental improvements that will deliver value to stakeholders. This allows making key decisions on laying out the facilities in early stages of a project rather than waiting for design to develop to a significant extent. Often times, project team wait too long for design to mature enough before applying CA/QRA to validate consequence impact and/or risk level. Most often, especially for Mega Projects, this leads to delays in executing the CA/QRA work properly at the right time to influence site layout. This delay in execution leads to scenarios when scope changes cannot be incorporated anymore. This translates into value erosion as project was unable to reap the full benefit of Facility Siting. The management framework recommended in this paper i.e. Change driven/adaptive framework allows for the stakeholders to start early in the project lifecycle to apply CA/QRA in laying out the plant facilities. The key is to start EARLY but START with the END in mind. Alignment among stakeholder on a VISION up-front is also important as this is expected to save life while striking a balance between business needs and climbing the Risk Mountain. Vision also allows avoiding scope creep while applying Adaptive/Change driven approach. It is highly recommended that stakeholders focus more on accuracy over precision. In the early stages of project lifecycle, quality information is scarce. Change driven/Adaptive framework gives you as the project stakeholder the opportunity to use "what you know" intelligently to ensure "what you don't know" falls into the right place.

6. References

[1] M. Mannan, A. Chowdhury and O. Reyes-Valdes, "A portrait of process safety: From its start to present day," College Station, 2012.

[2] M. Mannan, Lees' Loss Prevention in the Process Industries, Volumes 1–3 (3rd Edition), , 2005, Elsevier, 2005.

[3] K. McAndrews, "Consequences of Macondo: A Summary of Recently Proposed and Enacted Changes to US Offshore Drilling Safety and Environmental Regulation," in *Society of Petroleum Engineers, Americas E&P Health, Safety, Security and Environmental Conference*, 2012.

[4] E. W. Merrow, Industrial Megaprojects Concepts, Strategies and Practices for Success, New Jersey.: John Wiley & Sons, Inc., 2011.

[5] A Guide to the Project Management Body of Knowledge (PMBOK® Guide) – Fifth Edition, Newtown Square: Project Management Institute, Inc., 2013.

[6] S. Watson, "The Meaning of Probability in Probabilistic Risk Assessment," *Reliability Engineering and System Safety,* vol. 45, pp. 261-269, 1994.

348480 Global Energy and Transportation and Mobil Oil

Abstract:

Biofuels, Computational Methods, Engineering Economics, Polymers, Thermodynamics. Polymers-plastic, Thermodynamics- oil, Engineering Economics – the money it takes to finish the method of Global Engineering Research, Computational Methods – numerical input of analytical data, Biofuels – the transport fluids that utilized in more than one method from industrial usage to everyday living which enters the atmosphere. Exxon has completed a $600 million dollar deal for production of algae biofuel, amongst Mobil Oil has performed major research Global Energy and the effects of oil. Reading a most recent article on Mobil website.

348748 Characterization of Iron Phthalocyanine As the Active Material for Lithium Batteries

Abstract:

Results on the determination of the lithium-ion diffusion coefficient in the iron phthalocyanine (FePc) active material as a function of intercalation level, using a modified electrochemical impedance spectroscopy technique, will be presented. Also, developed thermodynamic functions such as the Gibbs free energy, entropy, and enthalpy changes for the lithium insertion reaction with FePc, derived from cell open-circuit voltage measurements as a function of intercalation level, are presented. Also presented are estimates of the maximum amount of lithium that can be intercalated into the FePc based on constant volume and temperature conditions.

349086 Design of a Free-Fall Reactor for Fast Pyrolysis of Waste Plastics

Abstract:

Thermal pyrolysis has earned significant reputation in the recovery of valuable chemicals, especially petrochemical based fuels from various materials, including biomass and plastics. Of many available technologies, *fast* pyrolysis was preferred in his study owing to its minimal heat and mass transfer limitations and better control on process conditions. The focus of this study is towards efficient conversion of waste plastics into gaseous products consisting primarily of alkanes, alkenes and aromatics. Lab scale experiments were performed using a micro reactor to determine favorable conditions, including reaction temperature and particle size that would result in maximum yield of ethylene. Using these results, a simple industrial size radiatively heated free-fall reactor was designed for the purpose of pyrolyzing waste plastics. The goal is to design a *fast* pyrolysis reactor that improves upon inefficiencies in traditional reactors while resulting in high yield of ethylene.

349732 Biosorptive Dehydration of Ethanol/Water Azeotropes Using Compound Starch-Based Adsorbent

Abstract:

Abstract: Recently, the interest in the production of bioethanol has greatly increased as a result of the increasing fossil fuel prices and associated environmental issues. Bioethanol is made from either grains or biomass and the broth contains 6 to 12 percent ethanol. In addition, water removal represents a serious problem as ethanol form azeotrope with water at atmospheric pressure. Distillation to water-free ethanol consumes 50 to 80 percent of energy used in the fermentation ethanol manufacturing process. Various techniques have been developed to break the azeotropic point, such as salting-out method, azeotropic distillation, extractive distillation, reactive distillation, adsorptive distillation and pervaporation. Among the methods, the adsorptive option is the most attractive, especially the biosorption, which has distinct advantages over conventional methods as it is environmentally friendly, highly selective, easily available, easy to operate, cost effective and reusable in repeated cycles in the treatment of aqueous organics.

In this work, a specially formulated compound starch-based adsorbent (CSA), which consists of corn, sweet potato and foaming agent, was developed. As an exploratory study, this research was focused on the investigation of CSA for biosorptive dehydration of aqueous ethanol and optimization of conditions for its maximum adsorption under various conditions of vapor feed flow rate, bed temperature and bed height through an orthogonal design method. Furthermore, to investigate the adsorption mechanism, dynamic saturated adsorbance of CSA was studied and characterization of the adsorbent before and after adsorption, as well as after regeneration, was performed.

In the first step, the net retention time and separation factor of ethanol and water were obtained using inverse gas chromatography (IGC), the results indicated that the net retention time of water on CSA is significantly longer than that of ethanol which attested the feasibility of this adsorptive separation purpose and low temperature could be propitious to the process which was consistent with the traditional theory of adsorption.

Then, through a batch adsorption experiment, the optimum adsorption condition was determined under vapor feed flow rate, bed temperature and bed height by orthogonal design method, which indicated that the optimization condition for ethanol was at vapor feed flow rate of 1.1g/min, bed temperature of 83 and bed height of 25cm. The packing density was 445kg/m^3. The residence time is determined by vapor feed flow rate and bed height. The residence time is more influenced than bed temperature. The longer the residence time, the more the output is. But it cannot be ignored that the adsorption selectivity may decreases. Low bed temperature is benefited to the adsorption process which agrees with the result of IGC. Five cycles experiments of adsorption-regeneration were carried out at the optimum adsorption and regeneration condition, the results showed that there is practically no change in the regeneration efficiency of CSA after the cyclic process adsorption–regeneration. It is due to the properly chosen regeneration conditions assuring complete desorption of the water adsorbed without changes in the CSA structure. This unchanged adsorption capacity after a number of cycles demonstrates reusability of the adsorbent.

To study the adsorption mechanism, experiment was conducted to study the dynamic saturated adsorbance of CSA. The water adsorption capacity of CSA has been found to be similar to some molecular sieves, which is about 0.15g/g. Besides, field emission scanning electron microscopy (FESEM) and mercury porosimetry (MP) were applied to investigate the change of surface morphology and microstructure of CSA before and after the adsorption, as well as after regeneration, as shown in Fig.1 and Fig.2. The FESEM images showed that compared with the particles before the adsorption, they became plump and the void spaces became smaller after adsorption, which indicated that the CSA was saturated with water. However, the CSA particles after regeneration became wizened like the nature particles, which demonstrated that almost all the water had been desorbed from CSA. The pore size distribution (PSD) indicated that compared with the PSD curve before adsorption, the peak at 6000nm disappeared and the peak at 17000nm became remarkably smaller after adsorption, which manifested that CSA is mainly composed of macropores which play a major role in the adsorption. However, the peaks at 6000 and 17000nm reappeared after regeneration, and the curve looked exactly the same with that before adsorption, which further demonstrated that CSA after regeneration can be totally regenerated and recovered to the original structure.

These results indicated that the starch-based adsorbent can be considered as an effective and low-cost bio-sorbent to dehydrate alcohol.

349781 Purification of 2-Amino-1-Phenylethanol Enantiomers By a New Technique Combining Distillation and Crystallization

Abstract:

Stripping crystallization (SC) is a new separation technique which combines distillation and crystallization. Basically, SC is operated at a triple-point condition, in which the liquid is simultaneously vaporized and crystallized due to the three-phase equilibrium. In essence, the process is continued until the liquid phase is completely eliminated and only the pure solid crystals remain in the feed. Thus, crystal washing is not required since only pure solid crystals remain in the feed and no impurities are adhered on the crystal surfaces at the end of SC. In the previous research, SC has been successfully applied to separate p-xylene from the mixed xylenes by Shiau et al. (2005, 2008).

Phenylethanolamine, also known as 2-amino-1-phenylethanol (2-APE), is an important basic skeleton in many adrenergic drugs and is often used as ingredients of cold medicines and anorectics. As the biological activity and pharmacokinetics of the enantiomers of a chiral drug can be different, the objective of this research is to apply SC to purify R-2-APE from the chiral 2-APE mixture. The experiments show that no crystallization occurs during SC without seeding due to subcooling of the liquid enantiomers. On the other hand, seeding can eliminate subcooling during SC, leading to the R-2-APE crystals gradually formed preferentially on the seeds. Thus, seeding with magnetic stirring can be successfully employed to purify R-2-APE from the 2-APE enantiomers.

1. Shiau, L. D., Wen, C. C. and Lin, B. S. Separation and purification of p-xylene from the mixture of m-xylene and p-xylene by distillative freezing, *Ind. Eng. Chem. Res*, **44(7)**, 2258-2265, 2005.

2. Shiau, L. D., Wen, C. C. and Lin, B. S. Separation of p-xylene from the multicomponent xylene system by stripping

crystallization, *AIChE Journal*, **54(1)**, 337-342, 2008.

350841 Fabrication of Low Cost Insulating Material from Kaolin Clay for Construction Purposes

Abstract:

A new low cost, thermal insulator, construction mineral polymeric material was synthesized from local Saudi Arabia resources using Inorganic Mineral Polymerization technique (MiP). In this technique, the mineral polymer is produced from the following components: clay mineral rich in koalinite, sand, sodium hydroxide solution, and water. Inorganic polymerization experiments are conducted by mixing known amounts of clay minerals and sand in a variable speed stainless-steel mixer for some period of times until a homogenous solid mixture is obtained. Sodium hydroxide solution is added to the mixture and the polymerization reaction takes place at different mixing speeds and times. The produced mixture is divided into a number of samples of known weights. Each sample is moulded in a stainless-steel cylinder at a medium pressure. Then the moulded samples are cured in an oven for 24 hours. The curing temperature was 80°C. The mineral polymerization conditions such as the ratio of clay to sand filler, particle size of the clay, particle size of the sand, concentration and amount of sodium hydroxide, speed and time of mixing were optimized to obtain stable and hard mineral polymer. Compressive strength for different specimens of fabricated mineral polymers was tested under dry and wet conditions. Values of compressive strength with values ranging from 11.5 MPa to 27.9 MPa were obtained.

351141 Thermodynamic Study of Binary PAH (Anthracene + Phenanthrene) Solid Mixtures

Abstract:

Polycyclic aromatic hydrocarbons (PAHs) are common components of many materials, such as petroleum and various types of tars. They are generally present in mixtures, occurring both naturally and as byproducts of fuel processing operations. It is important to understand the thermodynamic properties of such mixtures in order to better understand and predict their behavior (i.e., fate and transport) in the environment and in industrial operations. The Knudsen effusion technique was combined with gas chromatography and differential scanning calorimetry to measure the vapor pressure, composition, and phase behavior of binary anthracene and phenanthrene mixtures at near ambient temperatures. Mixtures of anthracene and phenanthrene exhibited non-ideal mixture behavior: In mixtures with less than 80% phenanthrene, this more volatile component preferentially sublimed and the vapor pressures decreased as the mixtures became enriched in anthracene. The vapor pressure data do not follow Raoult's law (that is, ideal mixture) predictions. Mixtures with high phenanthrene content (> 80%) showed azeotrope like behavior, in that anthracene and phenanthrene sublimed at nearly constant vapor pressure and molecular ratio. These results confirm that the behavior of few-component PAH mixtures does not follow ideal mixture models. Hence, as tars and petroleum (and potentially other organic compound-containing mixtures) weather to few component mixtures over time, complicated, non-ideal phase behavior might be observed.

351160 How Confident Are You That Your Major Accident Risks Are Under Control?

Abstract:

Major accidents in the process industries including those at Texas City , Buncefield UK and in the Gulf of Mexico have highlighted the critical role played by senior managers and executives. Effective leadership is essential if companies are to develop a positive safety culture that remains constantly vigilant towards the risk of a major process safety incident. In the UK the HSE has made leadership a key priority for the high hazard industries and there is a cross-industry initiative to develop and deliver training focussed towards board members responsible for process safety performance.

This paper explores the leadership failings that contributed to recent major accidents, and the cost to both the company reputation and the individuals involved. Key principles are discussed that can help improve leadership in the major hazard industries. These include; attitudes and decisions of senior manager affecting the investment strategy and the safety culture of the organisation, reinforcing the importance of safety by personal example, thorough understanding of major accident hazards and key risk control systems, investigating process safety incident and near misses to find the underlying causes, developing world class safety management systems, and identifying weaknesses in these systems using targeted performance indicators.

The author bases this paper on the development of a training standard for 'Process Safety Leadership' involving a cross-industry 'expert panel'. Training of senior executives in compliance with this standard has been carried out from late 2011 and feedback from these courses and the key requirements for effective leadership will be discussed.

351236 Sil Determination: Shortcomings with the Use of LOPA

Abstract:

Proposed Topic Area: Managing Risks

For a number of years, industry has been trying to apply the principles outlined in IEC 61508. In the process sector, this has resulted in a focus on the requirements of IEC 61511 (2003) and ANSI/ISA 84.00.01 (2004). These are risk based standards. They involve the identification of potential hazardous events, the causes that could lead to each event and the necessary risk reduction for each event. Where the risk reduction involves an instrumented function (e.g. a trip or an interlock), the required performance of the instrumented function can described in terms of the average probability of failure on demand. In the standards, the performance for such instrumented functions is assigned to one of four Safety Integrity Levels. The assessment of what performance level is needed in respect to a specific hazardous event is referred to as SIL Determination. In the field of SIL Determination, IEC 61511 provides examples of a number of different approaches – these are shown in IEC 61511 Part 3. These include Event Trees, Risk Graphs, Safety Matrices, Fault Tree Analysis and Layer of Protection Analysis (LOPA).

In the first few years, following the publication of IEC 61508 Edition 1 in 2000, industry generally embraced the use of Risk Graphs as the method of choice. However, over the subsequent few years, many companies using this method found that it was (a) giving higher safety integrity level targets than really needed and (b) was less flexible than other methods and consequently more difficult to use in many real situations. This has resulted in a general move to other methods and in particular to the use of Layer of Protection Analysis. Indeed, LOPA may now be the predominant "Method of Choice" for SIL Determination. As a method, LOPA gives the impression of simplicity and appears quite straightforward. It almost invites anyone with a minimum of understanding of the method to be a practitioner.

However, there are a number of shortcomings in the use of the LOPA method that have become evident through its use across the globe.

This paper will review the application of LOPA and provide insights into what may be seen as the proper use of the method and its core principles. It will also highlight various shortcomings in use that have been found during expert reviews carried out of a number of existing studies. The paper will be of benefit both to those who are new to the LOPA method, but also of value to those who have been using the method for some time.

Shale Oil and Gas – Wellhead to Fuel

Dr. Lawrence Kremer - Baker Hughes
AIChE - Hilton New Orleans Riverside
April 1, 2014

14AIChE
Spring Meeting, New Orleans, LA

BAKER HUGHES

Outline Unconventional Shale Oil

- Properties of shale oils
- Problems
  - Transportation
  - Refining

14AIChE
Spring Meeting, New Orleans, LA

BAKER HUGHES

General Characteristics of Shale Oils

- API gravity typically varies 30-55+
- Very paraffinic
- Low sulfur
- Can have hydrogen sulfide and mercaptans
- Low TAN
- Very low asphaltenes
- Light crude with very little residual material
- Many samples show great quality variation

Factor	Bakken	WCS
API	39.7	21.6
S%	0.25	3.44
Light ends	2.6%	
LSR	3.5%	3%
Light Nap	12.4	3.0
Heavy Nap	9.7	5.0
Jet	10.5	10.0
LFO	27.6	12.0
Gasoil	27.2	37.0
VacTar	6.7	29.8

14 AIChE Spring Meeting, New Orleans, LA

BAKER HUGHES

Shale Oil Property Comparison

Parameter	Eagle Ford	Bakken
API	52	40.8
TAN (g KOH/g)	< 0.05	0.09
Sulfur (wt %)	<0.2	0.304
Asphaltene (wt %)	0.1	0.41
Resin (wt %)	1.6	4.95
Filterable Solids (PTB)	225	76

14 AIChE Spring Meeting, New Orleans, LA

BAKER HUGHES

Utica Shale Crude - Three Phase

14AIChE
Spring Meeting, New Orleans, LA

BAKER HUGHES

Eagle Ford Crude Samples

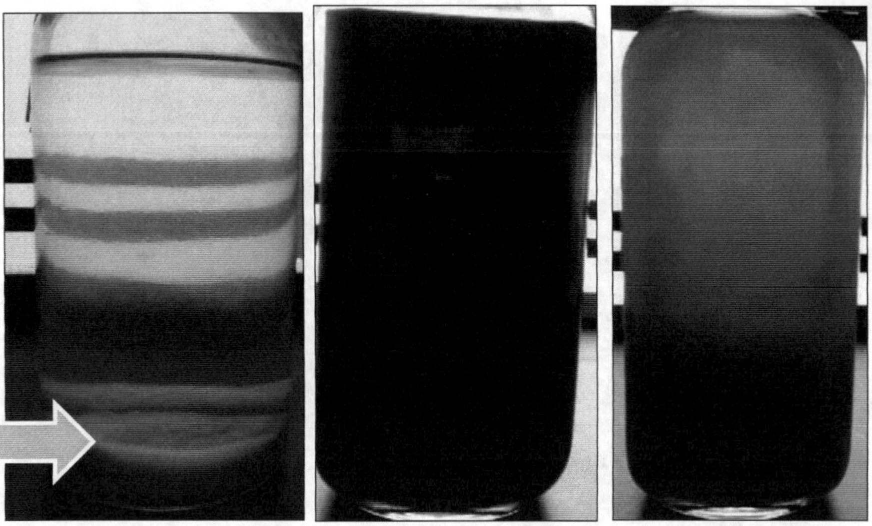

14AIChE
Spring Meeting, New Orleans, LA

BAKER HUGHES

Analytical Breakdown

Value	Yellow	Red	Black
API	55.0	44.6	52.3
TAN	<0.05	0.07	<0.05
Sulfur	<0.20%	<0.2%	<0.2%
Na	1.0 ppm	1.6 ppm	1.6 ppm
K	0.3 ppm	0.4 ppm	0.5 ppm
Mg	3.4 ppm	2.9 ppm	3.0 ppm
Ca	2.6 ppm	2.8 ppm	3.8 ppm
Asphaltene	0%	0%	0.1%
Resin	0.5%	3.2%	1.6%
Filterable solids	176 PTB	295 PTB	225 PTB
P	5.3 ppm	4.6 ppm	5.0 ppm

14 AIChE Spring Meeting, New Orleans, LA

BAKER HUGHES

Micrograph of Sludge

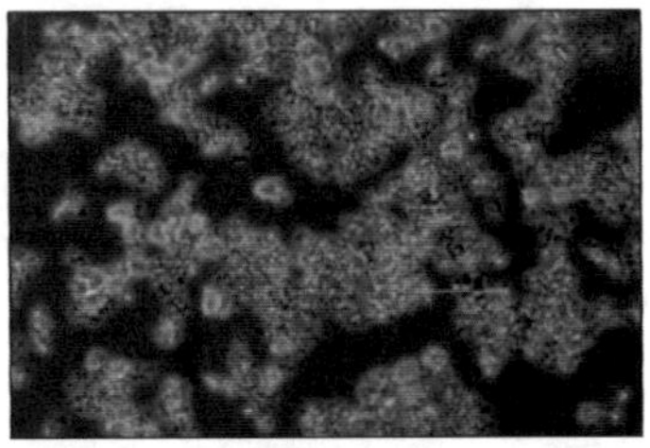

100 X

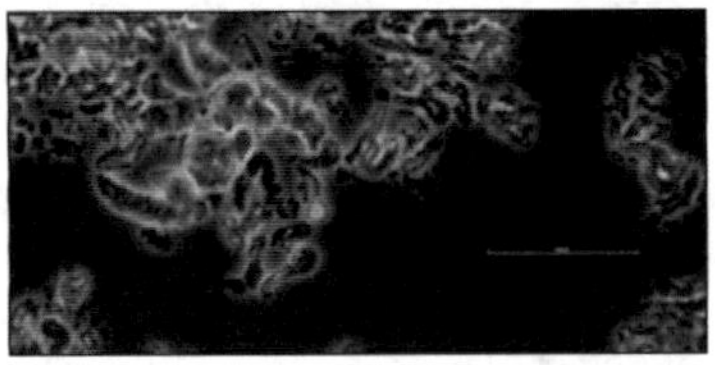

400 X

14 AIChE Spring Meeting, New Orleans, LA

BAKER HUGHES

Paraffin Analysis of Sample

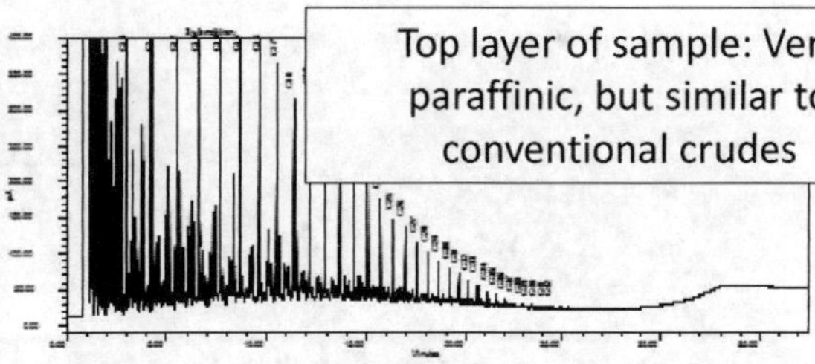

Top layer of sample: Very paraffinic, but similar to conventional crudes

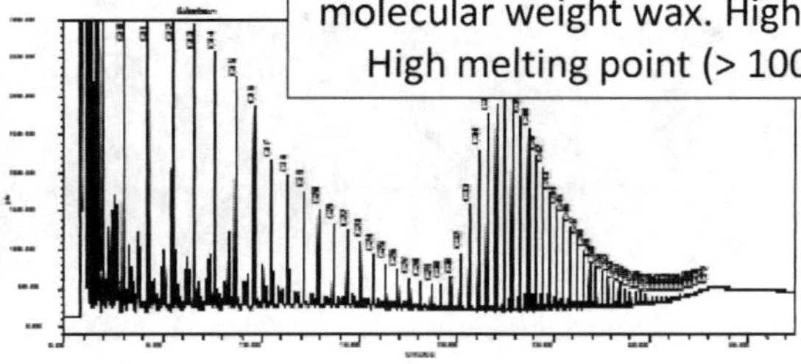

Bottom layer: Considerable high molecular weight wax. High MW = High melting point (> 100 °C)

14 AIChE Spring Meeting, New Orleans, LA

BAKER HUGHES

Transport of Shale Production

- Infrastructure does not exist to meet needs
  - Most shale oil moved by rail or trucks
  - Limited pipelines expanding to meet transportation needs
- Unit trains to East Coast, West Coast, and Gulf Coast
- Problems:
  - Tankers and barges with heavy wax deposits ROB
  - Tight emulsions that look like oil or white condensate but high BS&W
  - Cars not uniformly treated for H_2S
  - Corrosion in rail cars

14 AIChE Spring Meeting, New Orleans, LA

BAKER HUGHES

Viscosity of ROB Before Treatment

ROB from untreated compartment

14AIChE Spring Meeting, New Orleans, LA

BAKER HUGHES

Eagle Ford Shale Oil Wax Dispersion Testing

Several products were effective in dispersing the paraffin wax in an Eagle Ford shale oil sample at ambient temperature for over a month.

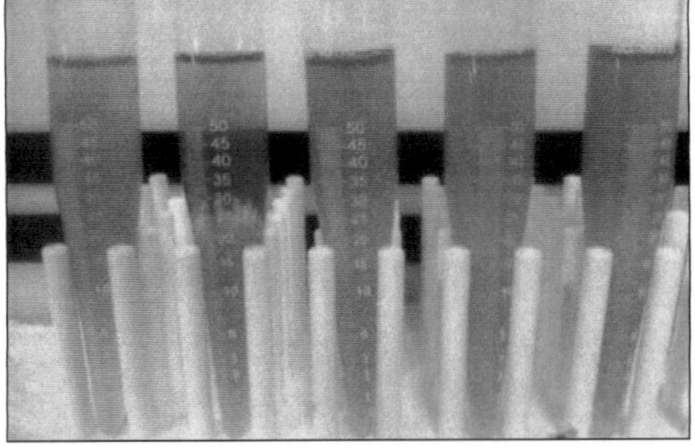

14AIChE Spring Meeting, New Orleans, LA

BAKER HUGHES

Product 'A' Effect on Wax Deposition (Utica)

Blank 200 ppm 400 ppm

14AIChE Spring Meeting, New Orleans, LA

BAKER HUGHES

Refinery Issues

- Crude Blending
- Fouling
- Corrosion
- Product quality

14AIChE Spring Meeting, New Orleans, LA

BAKER HUGHES

Crude Unit Fouling

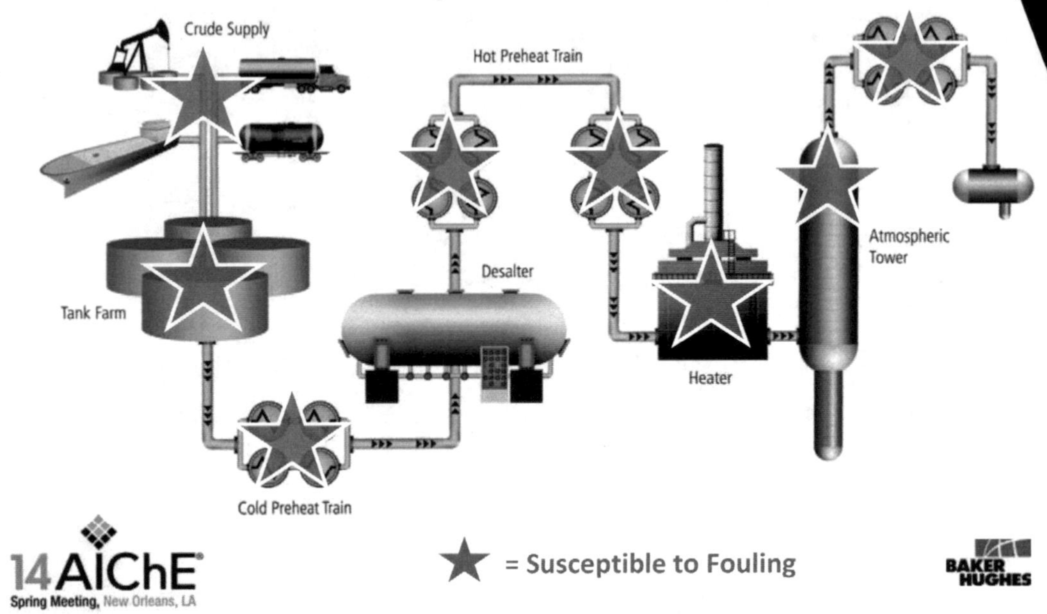

Asphaltene Precipitation

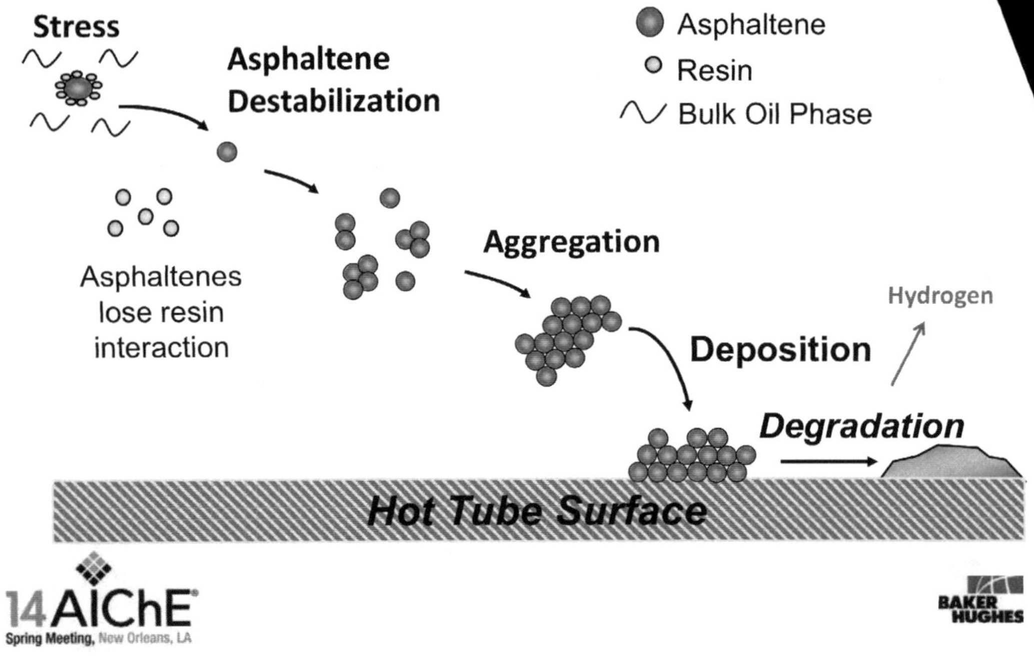

Asphaltene Stability Index Test

Field ASIT services™ technology measures the onset of the flocculation of the asphaltenes with high accuracy by inducing the asphaltene precipitation via titration with a paraffinic solvent.

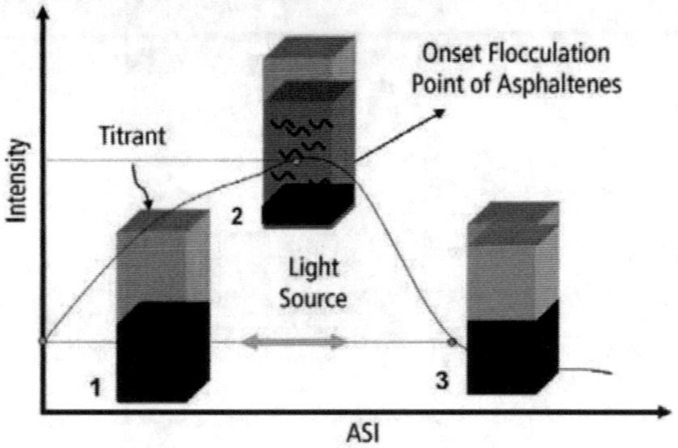

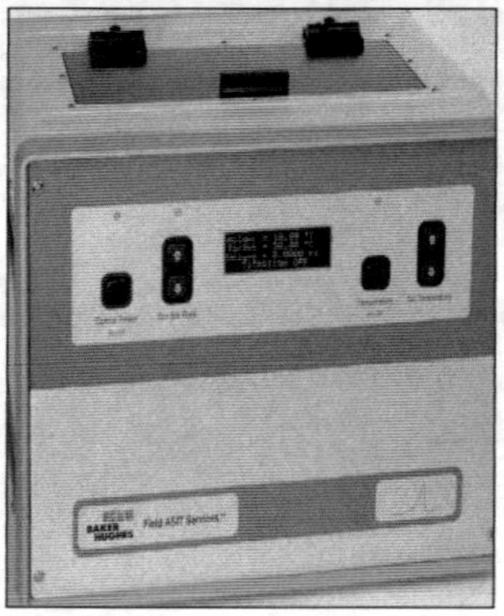

14AIChE® Spring Meeting, New Orleans, LA

BAKER HUGHES

ASI Shows Eagle Ford Can Destabilize Crude

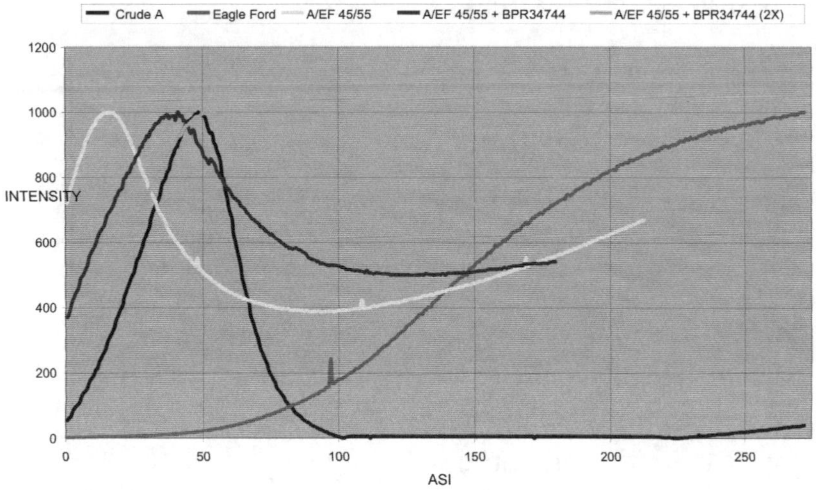

14AIChE® Spring Meeting, New Orleans, LA

BAKER HUGHES

Tower Tray Fouling - Phosphorus

- Seen by refineries running Bakken, Eagle Ford, and WTI
- Operationally manifested as increasing tower dP
  - Typically in the kero/jet zone
- Tower inspection shows deposits on trays in kero/jet zone
  - Some deposition may also show a few tray levels above and/or below kero/jet zone
  - Deposit scale typically tightly bonded to tray surface restricting flow and valve function

Above kero/jet zone

In kero/jet zone

Below kero/jet zone

14AIChE® Spring Meeting, New Orleans, LA

BAKER HUGHES

Shale Oil Impact – MEA Contamination

Corrosion and Fouling

- Shale oils transported by truck, rail, barge.
- Many shipments are treated with H_2S scavengers:

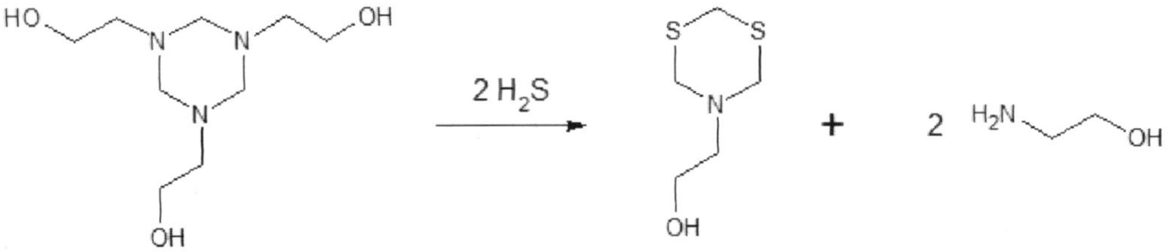

- Potential Impacts:
  - H_2S scavengers leave byproduct amine in crude oil (MEA, MA).
  - Tramp amines distill to OH system, create amine-HCl salts:
  - Corrosive to system metallurgy.

14AIChE® Spring Meeting, New Orleans, LA

BAKER HUGHES

Impact of Tramp MEA on OH Water

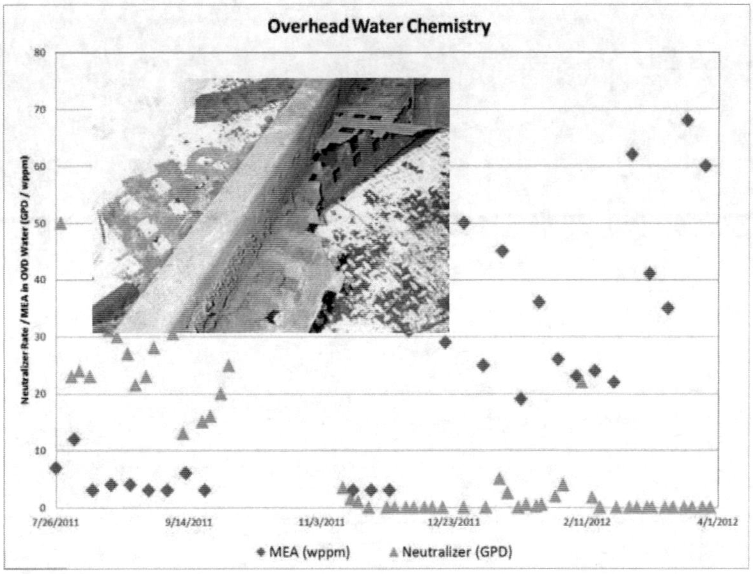

14 AIChE
Spring Meeting, New Orleans, LA

BAKER HUGHES

Salt Risk w/MEA Concentration

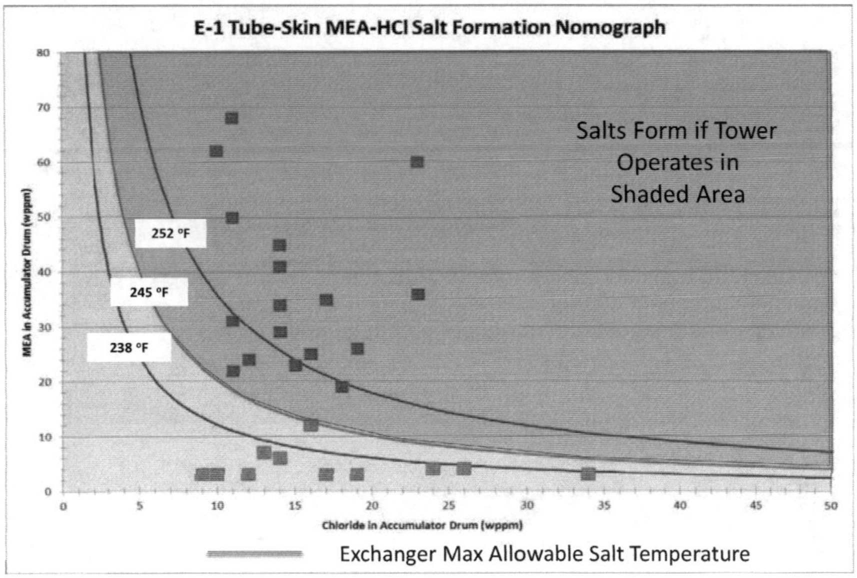

14 AIChE
Spring Meeting, New Orleans, LA

BAKER HUGHES

Shale Oil – Refinery Process Concerns and Monitoring

BAKER HUGHES

CHARACTERISTIC OF SHALE OIL	CONCERN IN REFINERY	PROCESS DATA	SAMPLES AND ANALYSES Samples	SAMPLES AND ANALYSES Analysis	OUTPUT DATA	FREQUENCY	BHI PROGRAMS
Crystalline Wax	Tank bottom sludge	• Tank Sludge Depth	• Tank bottom	• Visual sludge • GC analysis	• Tank Sludge Depth • Crystal Wax Presence	• Monthly	• XERIC™ Pretreatment
	Cold Preheat Fouling	• Desalter Temperature			• Temp vs. Time	• Update 2x/mo.	• LIFESPAN™ Wax Dispersants
	Desalter emulsions	• Emulsion layer growth	• Triline emulsions	• Visual wax • GC analysis	• Crystal Wax Presence	• After Shale Oil begins • When emulsion grows	• XERIC Demulsifiers
Asphaltene Stability in Blends	Desalter emulsions	• Emulsion layer growth	• Triline emulsions	• Precipitated Asphaltenes	• Precipitated Asphaltenes	• After Shale Oil beings • When emulsion grows	• XERIC Asphaltene Stabilizers • Field ASIT services™
	ColdPreheat Fouling	• Desalter temperature	• Crude blend	• ASIT	• Temp vs. Time • ASIT #	• Pre-Shale Oil • Update 2x/mo.	• LIFESPAN Antifoulants • Field ASIT services
	Hot Preheat Fouling	• Furnace inlet temp • Furnace tube temps • Furnace fuel usage • Preheat DP	• Crude blend	• ASIT	• Temp vs. Time • Fuel use vs. Time • ASIT #	• Pre-Shale Oil • Update 2x/mo.	• LIFESPAN Antifoulants • MILESTONE™ Furnace Fouling Control • Field ASIT services
Paraffinic	Furnace Coking	• Furnace tube temps • Furnace fuel usage	• Crude blend	• HLPS	• Coking tendency	• Pre-Shale Oil • W/Shale Oil	• MILESTONE Furnace Fouling Control
Tramp MEA	Tower/OH corrosion (Early Warning)	• OH Neutralizer rate • EXCAL usage rate	• OH water • Desalter brine	• OH pH • Brine pH	• Early Sign of Amine • Early sign of Amine	• Pre-Shale Oil • W/Shale Oil	• TOPGUARD™ Overhead Corrosion Control • EXCALIBUR™ Contaminant Removal Technology
	Tower/OH corrosion (Quantitative)	• Inputs for CRM	• OH water • Desalter ww • Desalter brine	• IC Amines	• Confirm MEA in Crude • OH MEA ppm • Salt Risk Tower • Salt Risk OH System	• Pre-Shale Oil • W/Shale Oil • Update 2x/mo.	• TOPGUARD Corrosion Control • EXCALIBUR Technology • TOPGUARD Corrosion Risk Monitor (CRM)
Phosphorus	Tower P Fouling	• Tower DP			• Tower DP Increase	• 2x/mo.	• LIFESPAN Antifoulants Dispersants
			• Crude	• ICP for P	• P in Crude	• 2x/mo.	
			• Kerosene	• Filt Solids • XRF Filter P • ICP for P	• Incr. Kero Solids • P in Solids • P in Kerosene	• 2x/mo.	
			• Kerosene • Diesel	• ASTM D86	• Incr. in Diesel IPB – Kero EP overlap	• 2x/mo.	

EXCALIBUR, FIELD ASIT SERVICES, LIFESPAN, MILESTONE, TOPGUARD and XERIC are trademarks of Baker Hughes Incorporated.

Development of a Fundamental Wax Deposition Model for Water-in-Oil Dispersed Flow

Mark Zheng and H. Scott Fogler

Wax Deposition in Off-shore Productions

2

- A challenge for flow assurance
  - Wax deposit at T_{wall} < WAT
- Significant costs related to remediation practice
  - Pigging
- Deposition modeling
  - Fundamental understanding
  - Enable effective remediation

wax deposition in subsea pipelines

cost as a function of pigging frequency

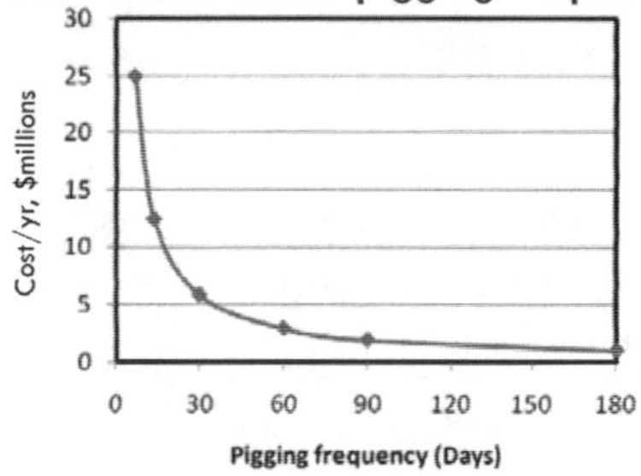

www.deepstar.org

Niesen, E&P Magazine 2002

The Michigan Wax Predictor (MWP) – the Most Advanced Wax Deposition Modeling Tool

3

- Developed through
  - 4 Ph.D. thesis at the University of Michigan over 15 years
- Verified by
  - flow-loop experiments from different facilities around the world
- The **only model** that solves the fundamental transport equations
- Predicts **consistent trends** of deposition rates with varying operating conditions
- Contains **no tuning parameter** for the upper and lower limits
- Contains **one tuning parameter** to account for wax precipitation in the bulk

Deposition Rate Predictions by The Michigan Wax Predictor (MWP)

4

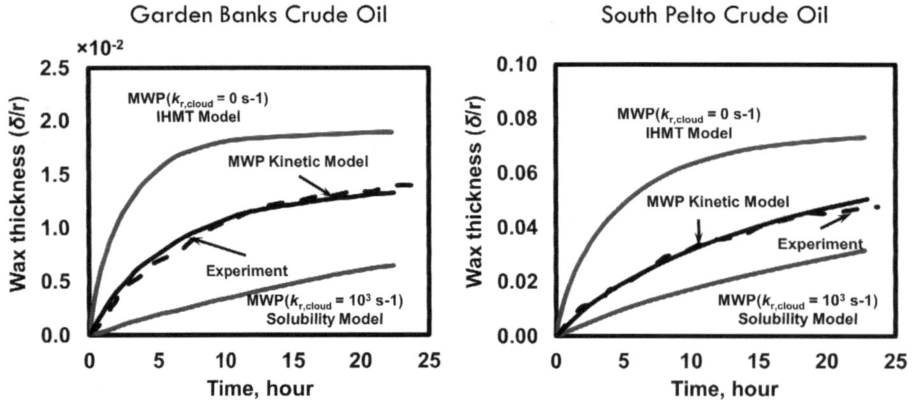

- Without tuning, MWP predicts upper and lower limits that bound the actual deposition rate.
- Predictions can match experimental deposition rate by tuning the bulk precipitation kinetic constant.

Huang et al. (2011)

More Deposition Rate Predictions by The Michigan Wax Predictor (MWP)

5

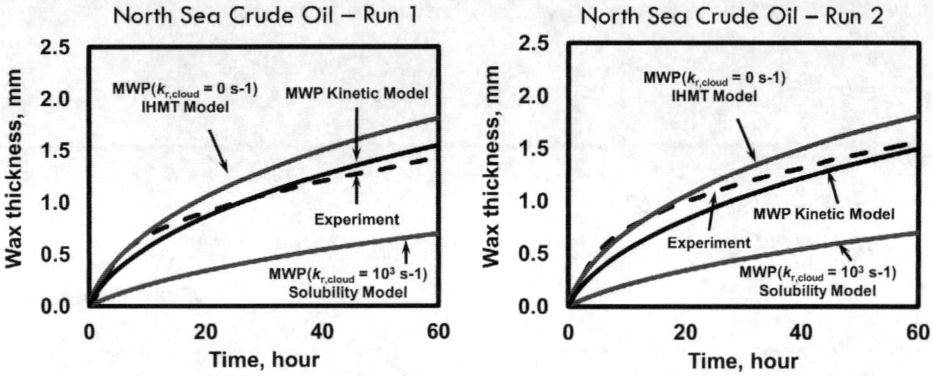

- Predicts consistent deposition rate with >20 more deposition experiments.

Huang et al. (2011)

Current Status of Wax Deposition Modeling in Water-in-Oil Dispersed Flow

6

- In order to model effects of water on the heat and mass transfer characteristics
  - Heat transfer: pseudo-single phase approach
  - Mass transfer: **no fundamental modeling approaches due to difficulties (see next slide)**
- Goal: based on MWP, build a fundamental wax deposition model for dispersed flow.

The Effect of Water Droplets on Mass Transfer: Partially Block the Diffusion Path

7

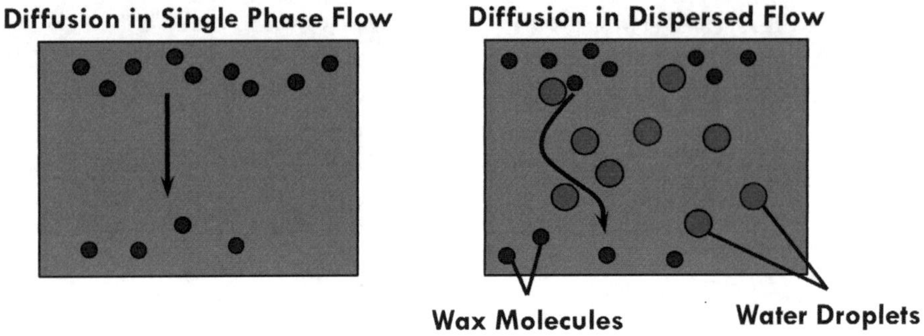

- Difficult to quantify the effect of water droplets on wax diffusion
- No established modeling strategies

Current Strategies for Modeling Diffusion in Water-in-Oil Dispersion

8

- Strategy 1: Use emulsion viscosity for diffusivity calculation
  - Does not represent the physics: diffusion occurs in the continuous phase
- Strategy 2: Use continuous phase diffusivity
  - Does not capture the effect of water droplets on diffusion.
- Neither approach is fundamentally correct

energy&fuels — Review — pubs.acs.org/EF

Review of Paraffin Deposition Research under Multiphase Flow Conditions

Cem Sarica* and Ekarit Panacharoensawad

It is still not clear if the mixture viscosity or the oil viscosity should be used for the wax diffusivity calculation in the Hayduk–Minhas equation[61] because this equation is not

Sarica et al. (2012)

Outline of the Project

9

- Investigation of diffusivity reduction due to water droplets
  - Verify
  - Quantify
  - Model

- Based on water-in-oil model emulsion

Preparation of Model Water-in-Oil Emulsion

10

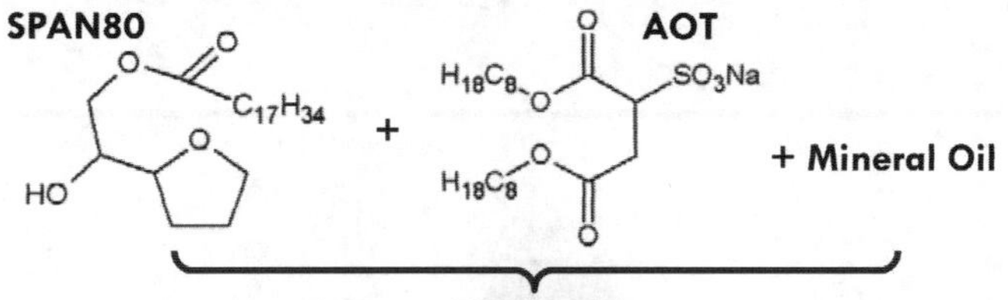

Surfactant:
5wt.% of total emulsion
(SPAN80: AOT = 9:1)

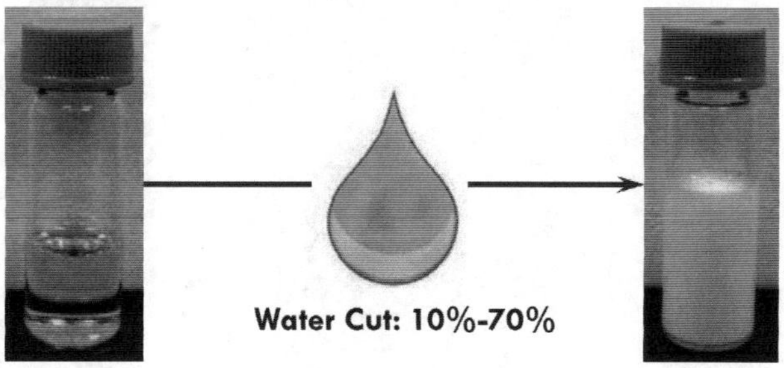

Delgado-Linares et al. (2013)

"Bulk Stability" of Emulsion

11

Time t = 0 hr

Time t = 48 hr

- Emulsion samples (Water cut = 10-70%) are stable over 48 hours.

Microstructure Stability – 1: Mean Droplet Size Constant Over 48 Hours

12

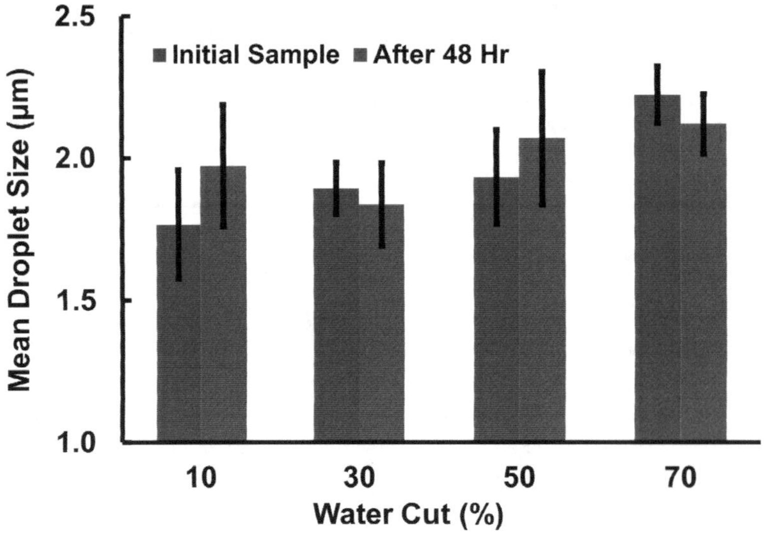

Microstructure Stability – 2: Droplet Size Distribution Constant Over 48 Hours

13

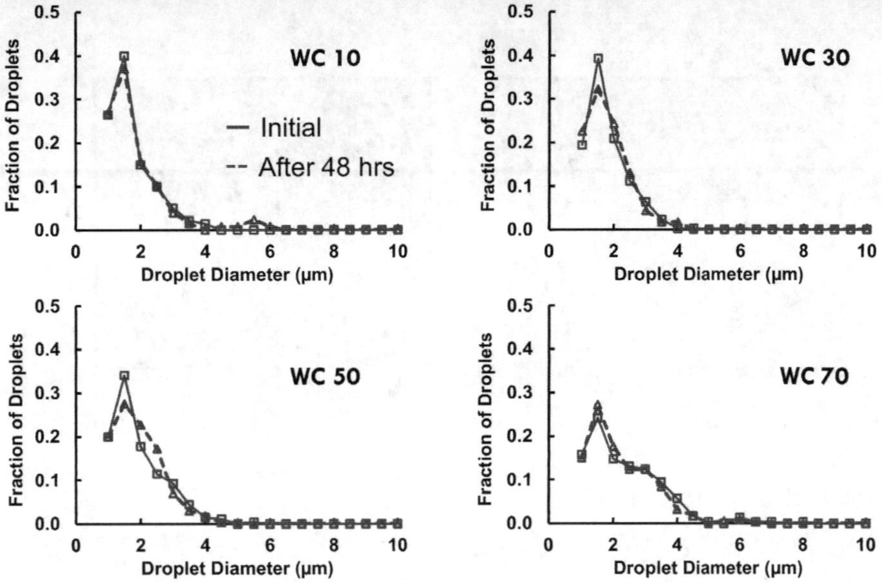

Summary

14

- Prepared emulsion with water cut ranging from 10 to 70% with 5wt.% surfactant
- Emulsions are stable for at least 48 hours
  - Macroscopically
  - Microscopically
- Can characterize the diffusivity of solute in emulsion
  - Study the effect of water drops on solute diffusion
  - Have a 48-hour window

The Idea of Using NMR to Characterize Diffusivity in Emulsion

15

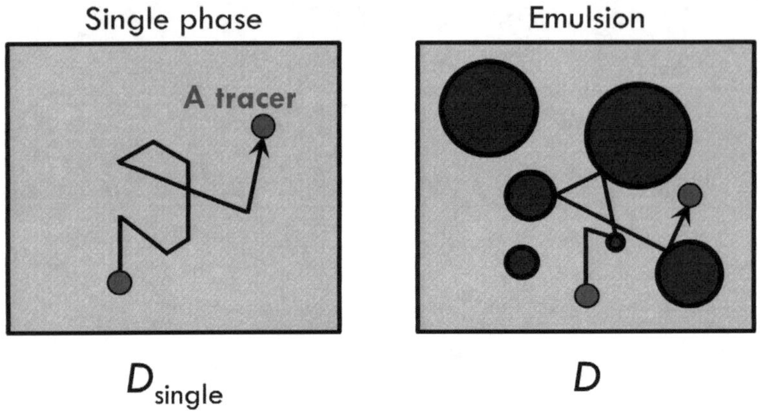

- Expected outcome: $D < D_{single}$
- Quantify the diffusivity reduction by calculating D/D_{single}

Identification of Tracer in Emulsion Mixture Using Its Chemical Shift

16

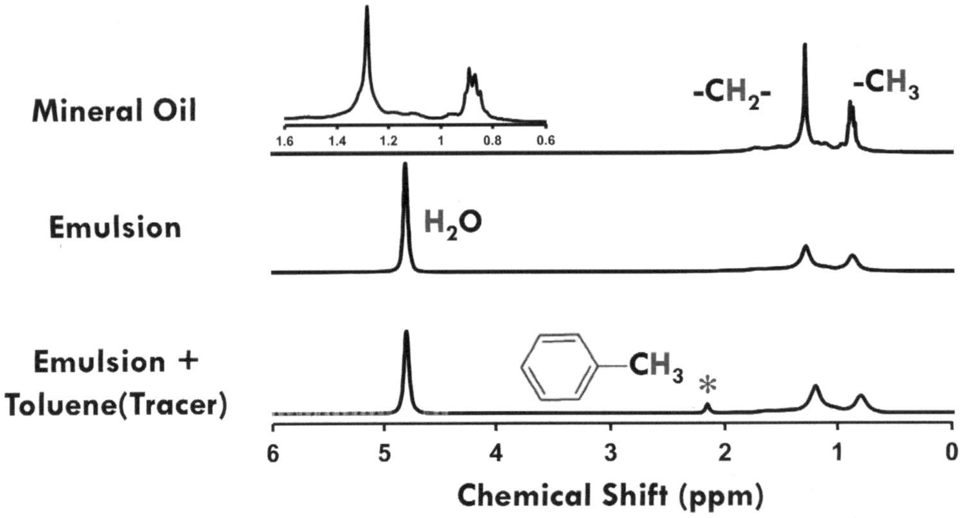

Characterization of Toluene Diffusivity in Emulsion

17

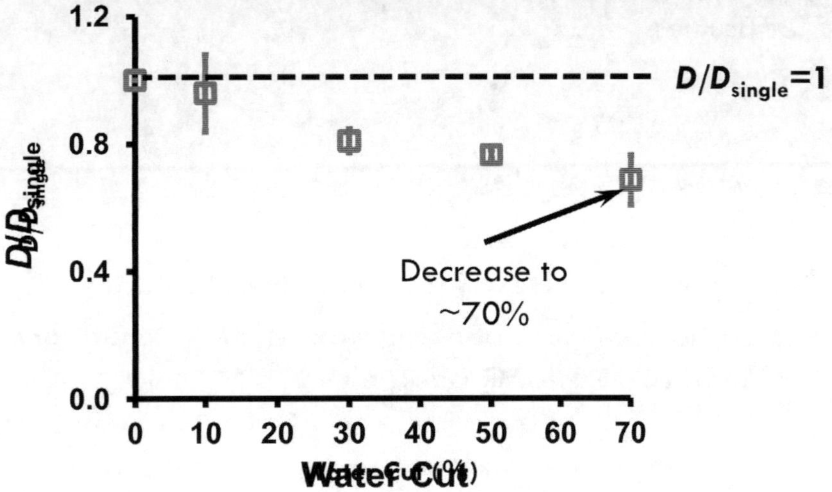

- At WC=70, diffusivity decreases to ~70% single phase diffusivity.

The Species Dependency of Diffusivity Reduce in Emulsion

18

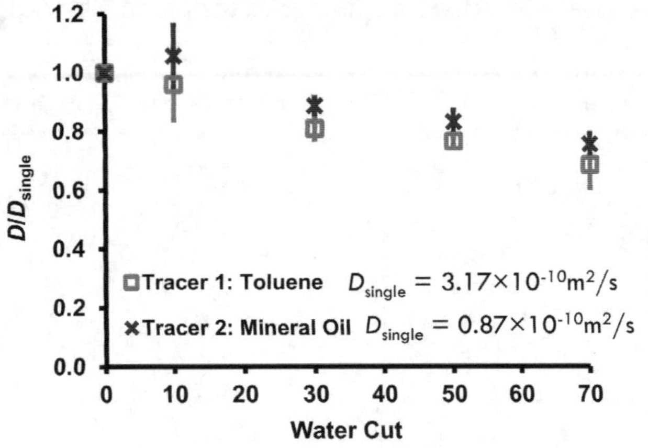

- The reduction in diffusivity seems to be "species-independent".
- Is it possible to measure wax diffusivity in emulsion?

Characterization of "Wax" Diffusivity in Emulsion – Part I. Difficulties

19

- Difficulties:

Wax, e.g. $C_{20}H_{42}$: 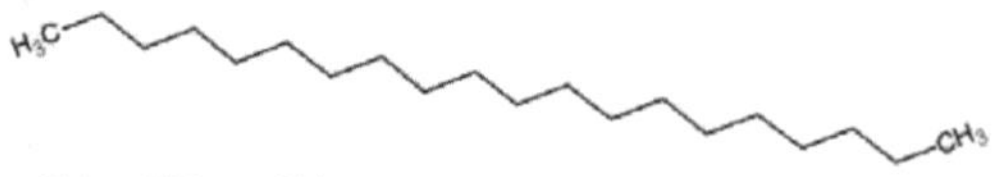

Mineral Oil: -CH, $-CH_2$, $-CH_3$

- Chemical structures of wax and mineral oil are similar
- Cannot distinguish between wax ^{1}H-NMR signal from the "mineral oil ^{1}H-NMR background"

Characterization of "Wax" Diffusivity in Emulsion – Part II. Solutions to Difficulties

20

- Use 1-fluorine-hexadecane to represent wax (F-Wax)

Regular wax:
cannot be detected in mineral oil medium

F-Wax:
can be detected in mineral oil medium

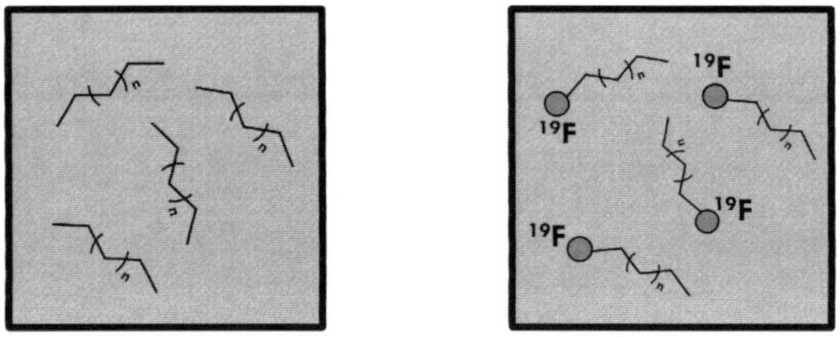

- "^{19}F tagged"
- Can track the diffusion of "F-Wax" by ^{19}F-NMR

Characterization of "Wax" Diffusivity in Emulsion – Part III. Results: Identification of F-Wax in ^{19}FNMR

21

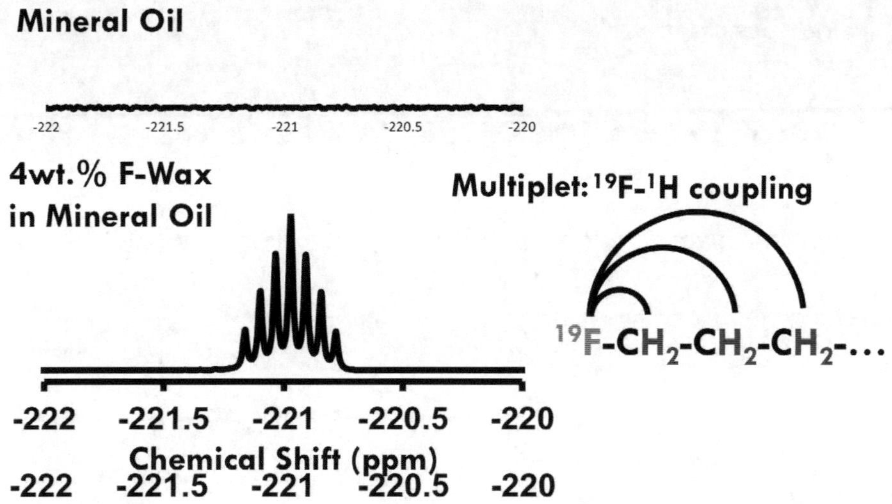

Characterization of "Wax" Diffusivity in Emulsion – Part III. Results

22

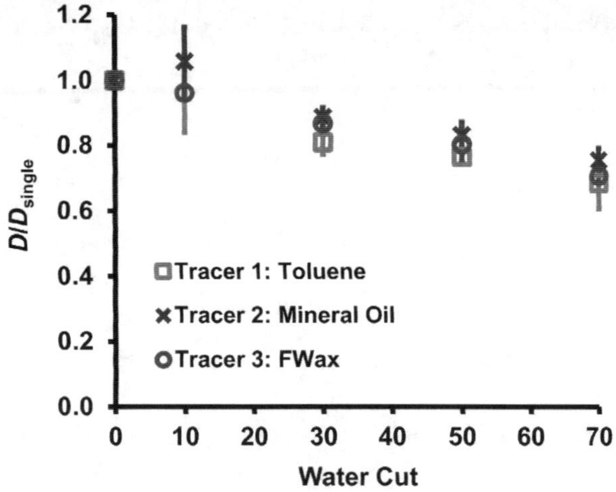

- Diffusivity reduction of a "wax-like" molecule and tracer molecules are similar.

Summary

23

- The reduction in diffusivity (D/D_{single}) due to droplets is species-independent.
- D decreases to ~70% D_{single} in WC70 emulsion
- Further investigation to model this diffusivity reduction

Strategy 1 for Diffusivity Calculation in Dispersed Flow

24

- Use emulsion viscosity in single phase diffusivity correlations

Hayduk-Minhas:

$$D \sim \frac{1}{\mu^{0.791-10.2/V_A}} \sim \frac{1}{\mu^{0.77}} \quad (V_A \sim 400 \text{ L/mol for wax})$$

Wilke-Chang:

$$D \sim \frac{1}{\mu}$$

Investigate the Effect of Using Emulsion Viscosity for Diffusivity Calculation

25

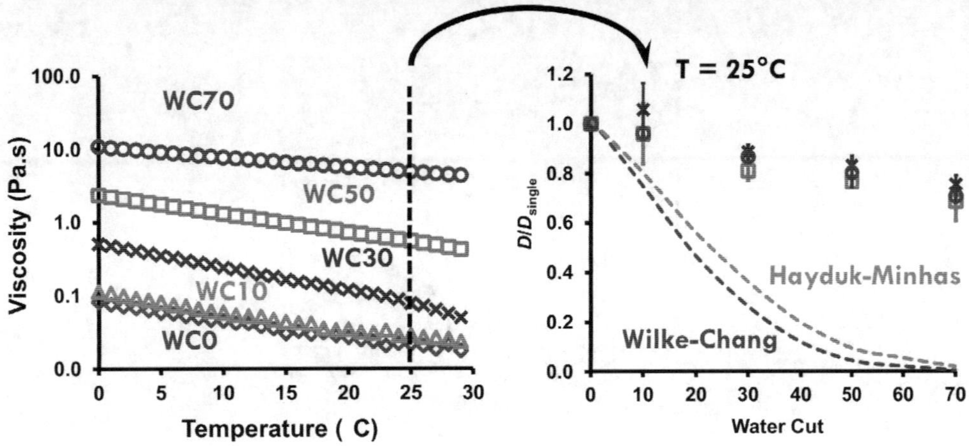

- Using emulsion viscosity to calculate diffusivity leads to 2 orders of magnitudes under-prediction.

Improved Strategy for Diffusivity Calculation in Dispersed Flow

26

Aris and Cussler (diffusion across membranes):

$$\frac{D}{D_{\text{single}}} = \frac{1}{1 + \frac{\alpha^2\Phi^2}{1-\Phi} + \frac{\alpha\Phi}{\sigma} + \frac{4\alpha\Phi}{\pi(1-\Phi)}\ln\frac{\pi\alpha^2\Phi}{\sigma(1-\Phi)}} \qquad \frac{D}{D_{\text{single}}} = \frac{1}{1 + \frac{\alpha^2\Phi^2}{1-\Phi}}$$

Jönsson (diffusion around spherical barriers):

$$\frac{D}{D_{\text{single}}} = \frac{2}{2+\Phi}$$

Aris et al. (1985), Cussler et al. (1988), Jönsson et al.(1986)

Predictions by Aris' Equation and Cussler's Equation

27

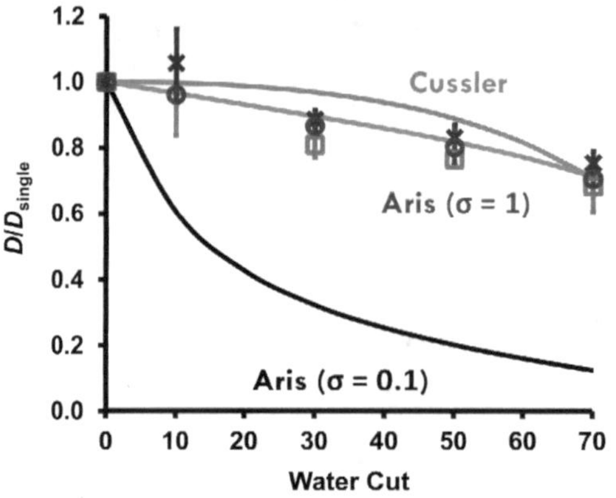

- Predict the correct order of magnitude
- Characterization of the spacing between droplets is difficult

Predictions by Jönsson's Equation

28

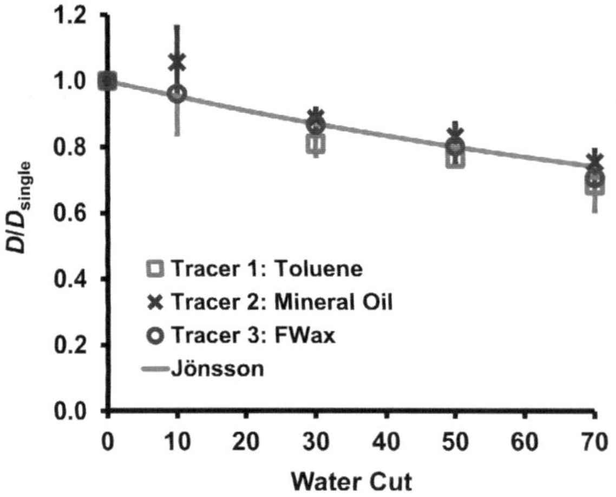

- Estimation by Jönsson's equation is in good agreement with experimental observations.

Summary

29

- Using emulsion viscosity to calculate the diffusivity causes two orders of magnitudes deviation in D/D_{single}.
- The effective diffusivity can be estimated based on the continuous diffusivity.
- Development and validation of the model in progress
  - Incorporation of dispersed phase diffusivity calculation into the Michigan Wax Predictor (MWP)
  - Validate by flow loop experiments

Acknowledgements

30

TOTAL
ConocoPhillips
Chevron
NALCO Champion
An Ecolab Company
msikenny
multi-chem
A HALLIBURTON SERVICE
HALLIBURTON
Statoil
bp

352143 Thermal Cracking Partial Upgrading Product Unstable and Incompatible

Abstract:

Bitumen and extra heavy oil containing asphaltenes are usually blended with condensate, light conventional oil, and, or fully upgraded, hydrotreated synthetic crude oil (SCO) for pipeline transportation and sale to the refinery market. Research and practice keep the ratio of light oil, condensate or SCO and the bitumen to the portions that will not cause the precipitation and deposition of asphaltenes. Some dispersants, flow improvers or additives are used in the bitumen blends. With blending as the most standard, common method of improving viscosity and density, there is little risk to this production transportation method.

Recently heated rail cars are used to transport bitumen without diluent. As the ratio of asphaltenes to other components such as resin does not change, there is little risk of precipitation of asphaltenes in this production transportation method.

During the period of high light crude price, low heavy crude price, high heavy-light crude differential, high diluent and high natural gas cost, partial upgrading schemes were proposed for in-situ down hole and at the wellhead. Typically, the partial upgrader proposals have not provided any flow assurance data that show asphaltenes do not precipitate or deposit in the process heaters, reactors, pipes, vessels, exchangers, downstream, or in storage.

Over twenty in-situ down hole methods are in various stages of pilot and testing techniques that might result in some partial upgrading. Over forty field located partial upgrading schemes have been proposed, none have been build or otherwise commercially proven. Down hole or above ground, the goal is to produce a lower viscosity material at the well-site, upstream, or midstream, that would require less diluent or natural gas. For example Fluid Catalytic Cracking, FCC, analogues UOP-CCU, Catalytic Crude Upgrading, Viscositor, Ivanhoe Energy IE-HTL, ETX-IYQ, NZ cracking use FCC configurations without hydrotreating the reaction product. For example, HTL uses high temperature short contact time to thermally crack bitumen vacuum bottoms.

Asphaltene deposition and diene gum formation are major flow assurance risks in partial upgrading thermal processing like HTL with high temperature short contact time reactions in an FCC analogue circulating sand.

The above ground field located partial upgrading methods have no commercial installations to validate performance. Despite claims of tens and hundreds of runs and commercial readiness, there is very little pilot plant or demonstration unit data on the product properties of the various methods; there is virtually no flow assurance data on the stability or compatibility of the various methods proposed for partial upgrading.

One thermal cracking partial upgrading process claims to have stable product, but is unable provide any data to demonstrate stability using any of the sixty possible test methods that have been used to show stability of fuel oils, crude, thermally and, or catalytically cracked products. On the other hand, problems associated with thermal reaction products are well known from the earliest days of refining with Shukhov, Burton and Dubbs process units. Unstable thermal reaction products at higher temperatures have been recognized and addressed throughout the long history of delayed, fluid and flexi coking, visbreaking, thermal and catalytic cracking.

High temperature thermal cracked reactions yield unstable products and with bitumen resid as feed most yield asphaltenes that are self-incompatible, which is completely different than hydrotreated, stable, and compatible SCO. Rather than overstate, confuse, or misrepresent the thermally degraded quality, partial upgraded material should be named differently from SCO, for example, Partially Upgraded Thermal Cracked Unstable Product would have acronym PUTCUP. PUTCUP with high olefins and dienes, disallowed by some pipelines, could be shipped like raw bitumen in rail cars, or if olefins allowed could be shipped like dilbit in pipelines or rail cars after blending with diluent to gravity or viscosity specification. PUTCUP includes heavy gum forming dienes in naphtha, jet and distillate fractions, self-incompatible asphaltenes and inorganic feed and produced solids in all fractions.

Patents and other open literature sources show short contact, high temperature reaction process products, like FCC, fluidcoker, HTL and pyrolysis naphtha and distillates, are less stable than lower temperature longer contact time products like visbreaker and delayed coker naphtha and distillates. These facts are in contrast to unsupported marketing claims, like HTL, of stable product due to shorter time. Simple chemistry and physics, it's about temperature, the higher the reaction temperature the more unstable product is formed, as shown by reaction product gases that have higher dienes including 1,3-butadiene. The higher the short contact reaction temperature, the more 1,3-butadiene in product gas with more liquid dienes throughout the reactor product, and the less stable the product. Thermal cracking processes with higher mix zone temperatures like HTL produce higher 1,3-butadiene, and corresponding higher C5-C10 dienes leading to high gum formation than process with lower reaction mix zone temperatures.

Processes, like HTL, with high temperature short contact time of fractions of seconds result in immediate reactor and downstream plugging with unstable self-incompatible asphaltenes reaction product precipitating and depositing with inorganic feed and produced solids and downstream heavy gum formation and deposits caused by dienes in naphtha, jet and distillate fractions. Thermal cracked product gum and residual asphaltenes deposit and plug reactors, transfer lines, quench towers, fractionation towers, downstream filters, catalyst guard beds, hydrotreaters, pumps, pipes and tanks. Self-incompatible PUTCUP is worse than just incompatible with other liquids as PUTCUP has precipitated, condensed and flocculated asphaltenes, semi-solids and solids that are out of solution, toluene insoluble, and cannot be resuspended or dissolved by any dispersant, flow improver, or other additive available.

No partial upgraders have been built to prove reliable, long term flow assurance of a year or more continuous operation, rather than a few hours of pilot or demonstration unit run before terminated due to plugging, etc.. Proof of reliable operation with industry standard on-stream factors over 95% uptime is necessary to reduce high risk of a high capital cost, ~ $550 million, investment in an unproven process. The ~$550 million for a 20,000 bpd partial HTL upgrader is like other partial upgrader estimates that typically don't include hydrotreating in order to show a significantly lower capital cost compared to full upgraders. However, partial upgrading reactor product is instantaneously self-incompatible with precipitated or condensed asphaltenes depositing and plugging in and on reactor walls, transfer lines, quench towers, pipes, and vessels downstream, and is inherently unstable due to dienes that form gums downstream of the reactor. PUTCUP fractions are thermally degraded compared to crude or bitumen. PUTCUP properties are similar to fluid coker product fractions.

Reduced flow assurance risk solutions to these partial upgrading challenges include the common practice of shipping bitumen in heated rail cars, or blending bitumen with diluent without partial upgrading.

Modeling Wax Deposition with Advanced Heat and Transfer Algorithms

(Jason) Zhenyu Huang, PhD
AIChE 2014 Spring Meeting

Assured Flow Solutions LLC

Introduction

Assured Flow Solutions LLC

- Wax deposition challenges
  - Wax deposit as Toil < WAT
  - A problem of $10^7
- Significant remediation costs
  - Pigging
  - Thermal insulation
  - Inhibitors
- Deposition modeling
  - Fundamental understanding
  - Enable effective remediation

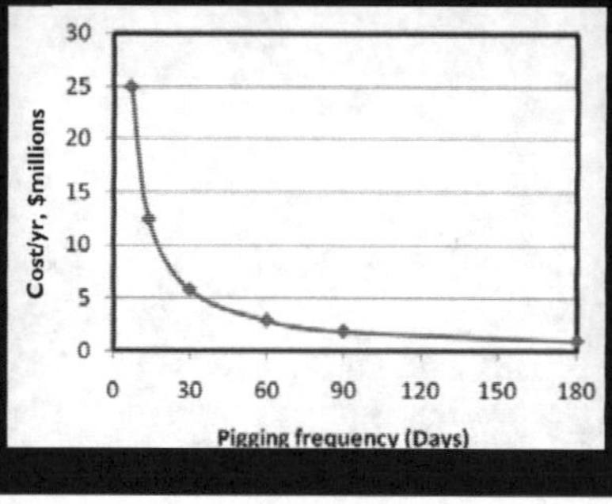

www.deepstar.org
Niesen, E&P Magazine 2002

Modeling Wax Deposition

Assured Flow Solutions LLC

Thermodynamics

Transport Phenomena

- How much does wax precipitate at a given temperature?
- How much does wax deposit at a particular location?

Wax Thermodynamics

Assured Flow Solutions LLC

- Waxy Components
  - n-C17+
- Wax Precipitation
  - WAT
  - Wax content
- Important Info to Further Investigate Wax Deposition Severity

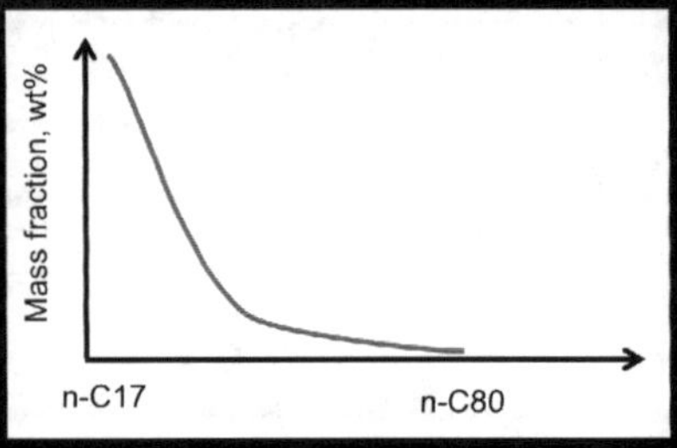

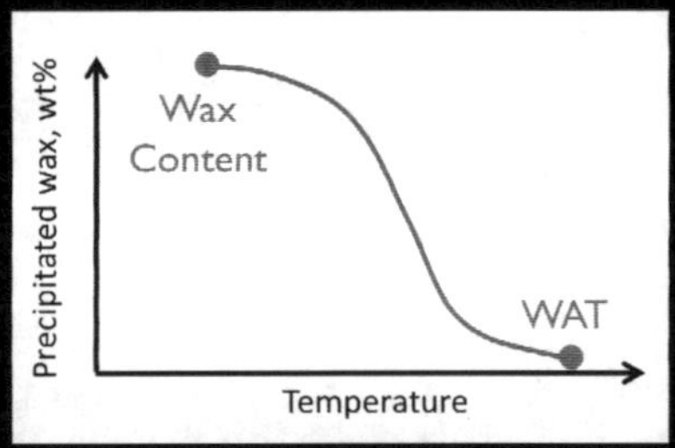

Understand Wax Thermodynamics

Assured Flow Solutions LLC

- Theoretical Modeling
  - Solid-liquid equilibrium of n-paraffins

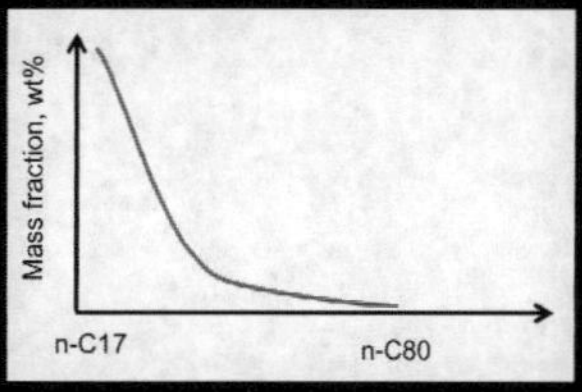

Equilibrium Types		PVTSim	GUTS	Multiflash
Vapor-Liquid Equilibrium	Vapor	Non-ideal, SRK Equation, (Won 1986)		
	Liquid			
Solid-Liquid Equilibrium	Solid	Single ideal solution	Multiple ideal solid solutions	Single or multiple non-ideal solid solution
	Liquid	Non-ideal (SRK)	Ideal solution	Non-ideal, Flory free volume theory (Coutinho, 1996)

Pedersen, K.S. SPE Production & Facilities, February 1995:46-49
D.D. Erickson, V.G. Niesen, and T.S. Brown, 1993, SPE 26604.
Coutinho, J.A.P., Stenby, E.H., Ind. Eng. Chem. Res. 1996;35:918

Understand Wax Thermodynamics

Assured Flow Solutions LLC

- Experimental Investigation
  - Cross-Polarized Microscopy (CPM)
    - Mostly for WAT
  - Differential Scanning Calorimetry (DSC)
    - WAT and precipitation curve
- Provide Wax Thermodynamics at stock tank conditions
- Important for Thermodynamic Model Tuning

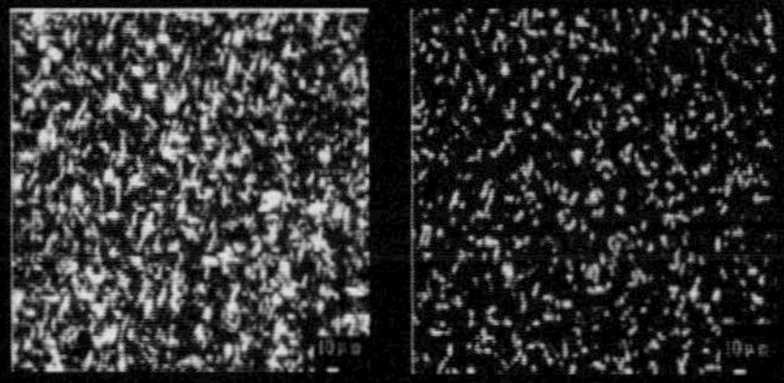

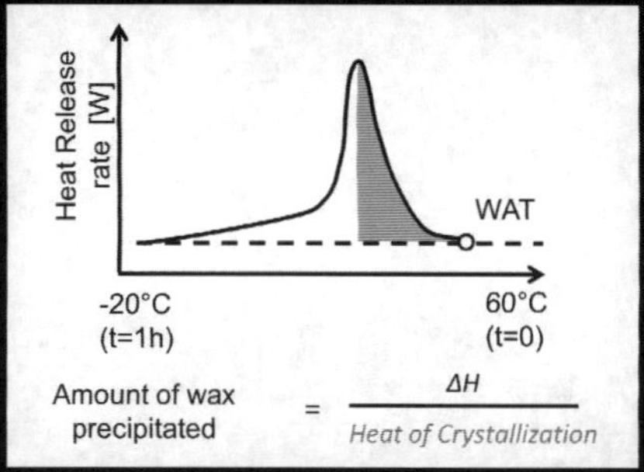

Pedersen, K.S. SPE Production & Facilities, February 1995:46-49
D.D. Erickson, V.G. Niesen, and T.S. Brown, 1993, SPE 26604.
Coutinho, J.A.P., Stenby, E.H., Ind. Eng. Chem. Res. 1996;35:918

Wax Deposition – Transport Phenomena

Assured Flow Solutions LLC

- Heat Transfer
  - Temperature distribution of the pipe
- Mass Transfer
  - What happens to the waxy components in the pipe?
- Fluid Dynamics
  - Gas/Water/Liquid flow

Studies on Wax Deposition Modeling

Assured Flow Solutions LLC

Earliest studies (1980s) (Understand Wax Deposition Mechanism)	Newberry, 1984 Majeed, 1990 Hsu, 1995
Studies around 2000 (Develop industrialized and academic models)	Rygg, 1998 (OLGA Wax) Matzain 2001 (OLGA Wax) Lindeloff and Krejberg 2002 (PVTSim DepoWax)
Recent academic studies (Understanding the transport phenomena in wax deposition)	Singh, 2001 Vekatesan 2004 Lee, 2008 Huang, 2010, 2011 Sheng, 2013 } The Michigan Wax Predictor

A gap seems to exist between industrial models and academic developments

Comparison on Wax Models

Assured Flow Solutions LLC

- Difficult - Lack of Good Data
  - Different apparatus were used
  - Need well-designed experiments covering a series of factors
    - (Oil, flow rate, temperature)
- Recent Study by Hoffmann et al.
  - 3 Years of experimental development
  - Indoor 2", 5.5m flow-loop
  - 50+ wax deposition experiments
  - Allow comprehensive model comparisons

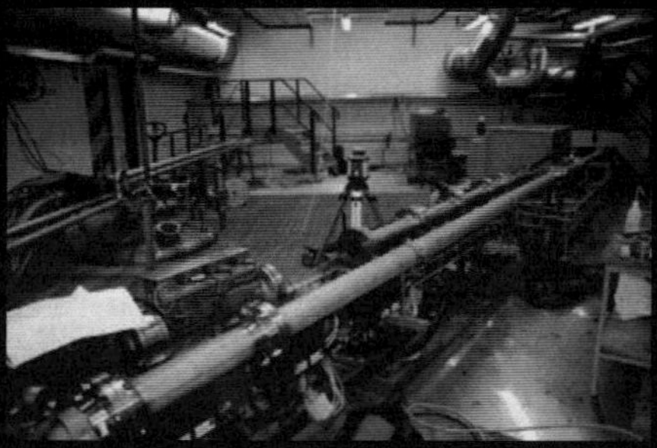

Evaluating Wax Deposition Models

Assured Flow Solutions LLC

- Compare Industrial and Academic Wax Models
  - OLGA Wax
  - MWP
- Oil studied
  - Waxy condensate from North Sea (30C WAT)

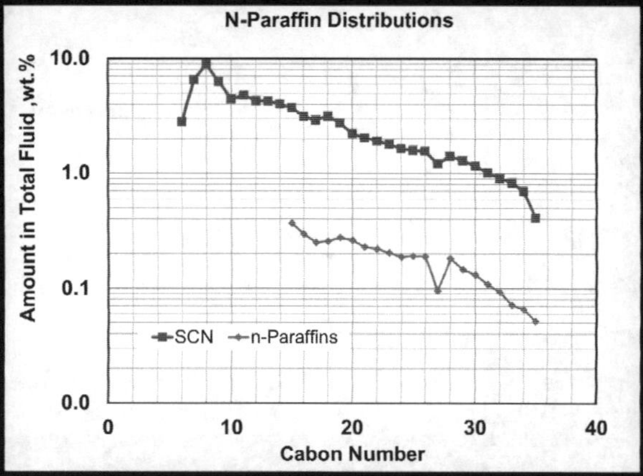

Input For Deposition Modeling

Assured Flow Solutions LLC

- Based on the same thermodynamics
  - Wax precipitation curve
  - Fluid viscosity

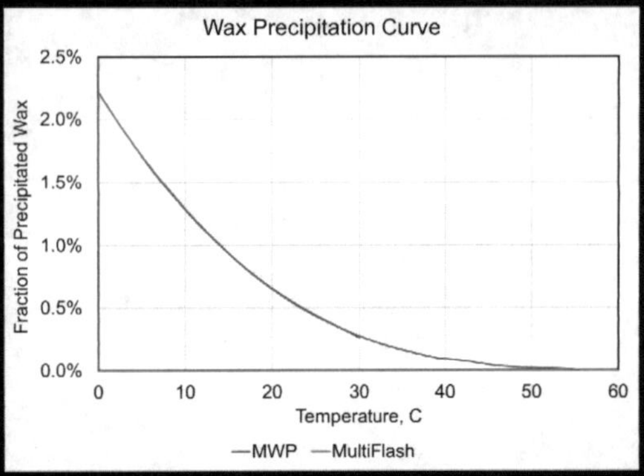

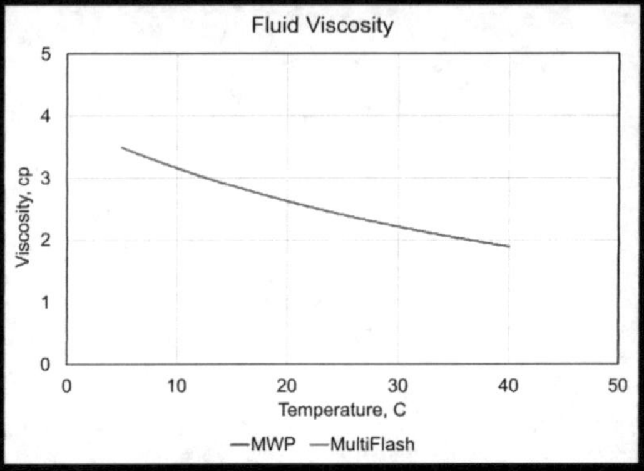

Results - Effect Of Oil Flow Rate

Assured Flow Solutions LLC

Experiment

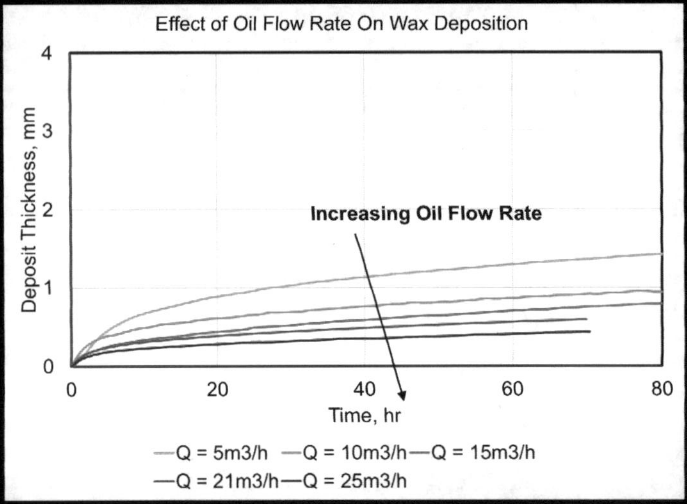

Deposit thickness decreases with increasing flow rate

Results - Effect Of Oil Flow Rate

Assured Flow Solutions LLC

Model Prediction (OLGA Wax)

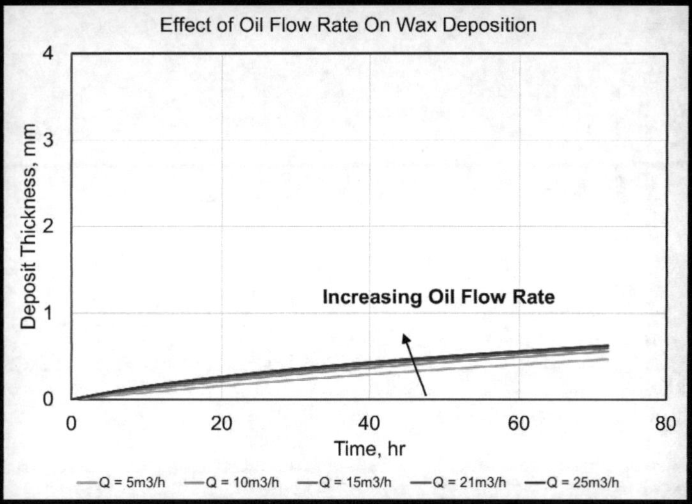

- Under-predicted in terms of Magnitude
- Trend opposite to experiments

Results - Effect Of Oil Flow Rate

Assured Flow Solutions LLC

Model Prediction (MWP)

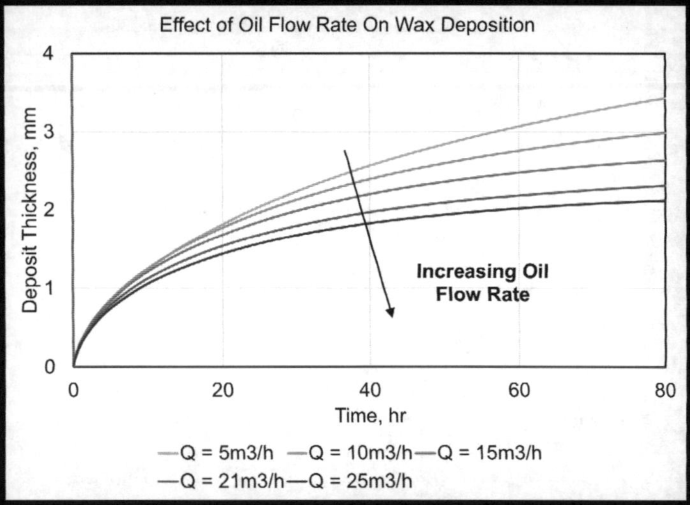

- Slightly over-predict in terms of magnitude
- Same trend as experiments

Results - Effect Of Coolant Temperature

Experiment

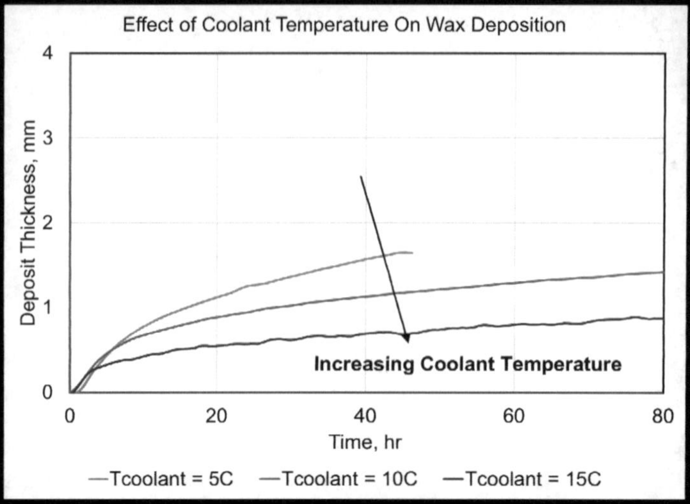

Deposit thickness decreases with increasing coolant temperature

Results - Effect Of Coolant Temperature

Model Prediction (OLGA Wax)

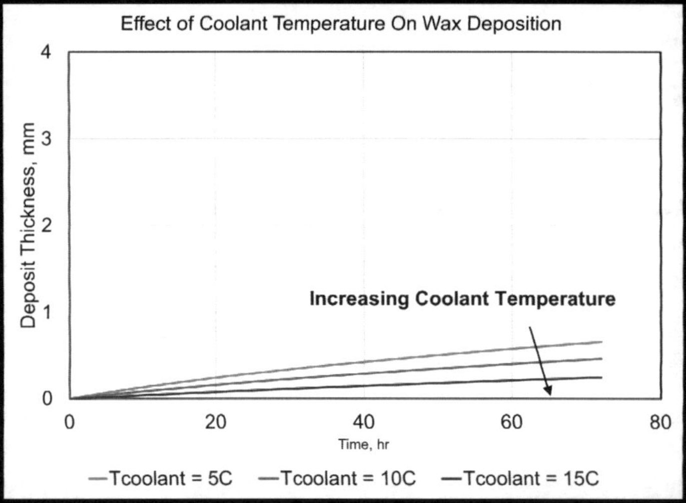

- Under-predicted in terms of Magnitude
- Trend same as experiments

Results - Effect Of Coolant Temperature

Model Prediction (OLGA Wax)

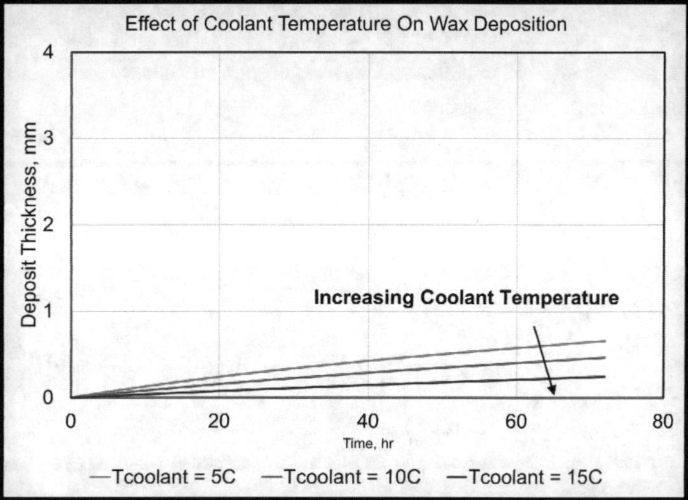

- Under-predicted in terms of Magnitude
- Trend same as experiments

Results - Effect Of Coolant Temperature

Model Prediction (MWP)

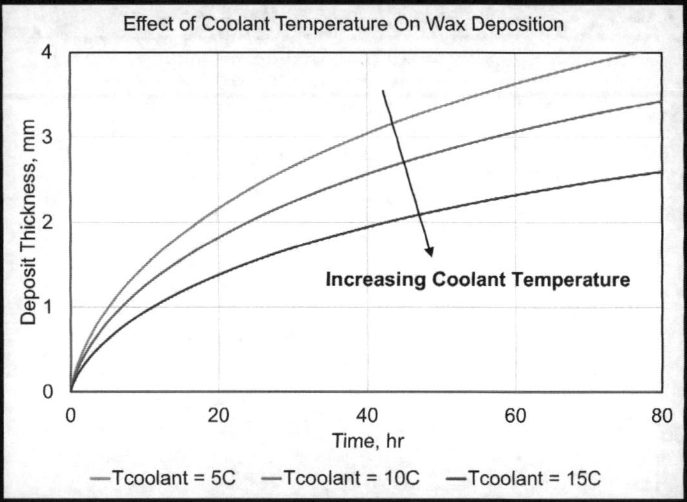

- Slightly over-predicted in terms of Magnitude
- Trend same as experiments

Summary

Assured Flow Solutions LLC

Operating Conditions	Experimental Trend	OLGA Wax	Michigan Wax Predictor
Increasing Flow Rate	Decreased Thickness	Increased Thickness	Decreased Thickness
Increased Coolant Temperature	Decreased Thickness	Decreased Thickness	Decreased Thickness

- Investigate the reasons for the difference:
  - Heat Transfer
    - OLGA Wax: 1D, MWP: 2D
  - Mass Transfer
    - MWP: Aging
    - OLGAWax: No Aging

Examination on Heat Transfer

Assured Flow Solutions LLC

- Pre-deposition Radial Temperature Profile

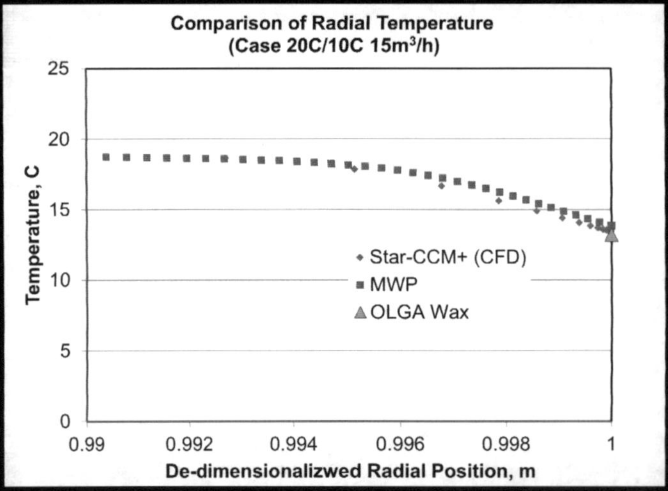

Temperature profile consistent with each other

Examination on Heat Transfer

Assured Flow Solutions LLC

- Pre-deposition Radial Heat Flux

Model	Star-CCM+ (CFD)	OLGA Wax	MWP
Heat Flux (W/m)	1009	928	944

Conclusion:

- Current industrial wax deposition models do not seem to accurately predicts the effect of operating conditions on wax deposition
- Heat transfer calculations seems consistent between three models
- The better performance in MWP is most likely due to its mass transfer calculations

Modeling Aging in Wax Deposition

Assured Flow Solutions LLC

- Aging: Increase of wax fraction in the deposit
  - 4% wax in oil, 30%-50% wax in deposit
  - Due to internal diffusion of wax in the deposit
- Why do we need to model aging?
  - It affects wax mass balance (corrected in some of the industrial models)
  - It affects diffusion of waxy components
  - It affects the impact of insulation by the deposit layer
- How do we model aging?
  - GC characterization of wax deposit
  - Understand the solid-liquid equilibrium of waxy components in the deposit (*Sheng 2013, Energy and Fuels*)

Assured Flow Solutions LLC

Thank You!

352711 Development of Experimental and Modeling Methods to Understand Asphaltene Precipitation

Abstract:

Asphaltene Precipitation and subsequent deposition during oil production is a potential problem in many oilfields around the world, especially for light crude oils with relatively low asphaltene content, or in oilfields that are treated with miscible gas injection.

Different experimental procedures and modeling methods have been proposed by several research groups trying to understand the mechanisms of asphaltene precipitation and deposition. However, because of the complexity of these phenomena, there is still active discussion about the theories governing asphaltene stability.

In this work we present experimental evidence that supports the theory that precipitation and aggregation of asphaltenes is a multi-step process, where the former is driven primarily by thermodynamics whereas the latter is driven by kinetics. Under this multi-step mechanism, asphaltene precipitation, which leads to the formation of an impure asphaltene rich phase in liquid-liquid equilibrium with an asphaltene-lean phase, is a fully reversible process. On the other hand, from the precipitated phase, subsequent aggregation and aging leads to the formation of more solid-like structures. Furthermore, we will present experimental results that suggest that the currently available commercial technologies to detect asphaltene precipitation (i.e. NIR spectroscopy and High Pressure Microscopy) might not be appropriate to detect the exact point of asphaltene precipitation, but instead they give a combined reading of precipitation plus aggregation. For this reason, the results obtained using standard Solid-Detection-System are extremely sensitive to the depressurization rates.

These experiments are in conceptual agreement with the modeling approach for asphaltene stability based on the PC-SAFT equation of State. It was already proven that the PC-SAFT EOS can provide very accurate predictions of the phase stability of asphaltenes, bubble point of the oil, density of gas and liquid phases, GOR, composition of liberated gas and even asphaltene compositional grading that can induce in extreme cases the formation of tar-mats. In this work, we will present the progress of an enhanced characterization method that has been applied to various wells from three different fields in the Middle East. We found that it is possible to reduce from three to one the number of adjustable parameters for asphaltenes and we are still able to get comparable results for the phase stability of asphaltenes as well as the other properties mentioned above. The new method simplifies the fitting process and reduces its computational time significantly, without sacrificing the accuracy of the predictions.

With these results we aim to contribute to the understanding of the complex phenomena associated to the asphaltene phase behavior and the prediction of the occurrence and the magnitude of asphaltene precipitation and deposition.

10TH GLOBAL CONGRESS ON PROCESS SAFETY

Piper Alpha 25 Years – Progress and Challenges Remaining for the Offshore Industry

Dr Robin Pitblado
DNV GL
1400 Ravello Dr, Katy TX 77449
robin.pitblado@dnvgl.com

Prepared for Presentation at
American Institute of Chemical Engineers
2014 Spring Meeting
10th Global Congress on Process Safety
New Orleans, LA
March 30 – April 2, 2014

UNPUBLISHED

Piper Alpha 25 Years – Progress and Challenges Remaining for the Offshore Industry

Dr Robin Pitblado
DNV GL
1400 Ravello Dr, Katy TX 77449
Robin.pitblado@dnvgl.com

Keywords: Piper Alpha, Process Safety, Progress

Abstract

The Piper Alpha disaster in July 1988 was a wake-up call to the upstream industry that simpler engineering and operational practices were no longer suitable for the challenging environments then being encountered in the North Sea. The industry has adopted many changes since then affecting offshore organization, process safety management, technical design of topsides, subsea and drilling systems, and human factors and safety culture. A major conference in Aberdeen, Piper 25, reviewed many of the positive changes that have occurred since that accident.

Benefits associated with safety cases in the UK and Australia are presented, along with improvements in safety management systems and safety culture programs. The many technical changes and improvements to fire and explosion knowledge and structural integrity are summarized. Progress on human factors has been less dramatic, but even here positive improvements are noted. Overall, there have been important improvements to process safety in the North Sea following Piper Alpha that would benefit all offshore regions,

1. Introduction

1.1 The Importance of the Piper Alpha Tragedy

Few accidents have had as much impact as Piper Alpha. The event showed how poorly the industry had prepared for a major accident of this type, and the escalation and number of fatalities was as unprecedented as it was unnecessary. The accident was followed by a major public inquiry and then the regulator and the industry worked hard to enhance offshore safety with many innovations.

1.2 Summary of Piper Alpha Event

Piper Alpha was an offshore major accident event in the UK and perhaps not familiar in detail to many US audiences. Some key facets of this accident are summarized here to provide some context for the safety developments which have occurred since.

Piper Alpha was a large fixed jacket facility in the UK North Sea with its own production and also acting as a distribution point for major oil and gas lines returning to shore via large riser pipes. In total it was processing around 10% of North Sea production at the time. It was located NE of Aberdeen in 474ft of water. A maintenance / operational error led to a release of hydrocarbons and a major explosion occurred on 6 July 1988, and this immediately impacted the control room and impaired the response. An ongoing pool fire led afterwards to the rupture of a major gas riser pipe and the event escalated dramatically. The crew mostly took shelter in the accommodation module but they never evacuated. Some crew jumped overboard and they were rescued. In total 165 men died from the platform plus additionally 2 rescuers, while there were 61 survivors. Good descriptions of the event are contained in Lees [1] and CCPS [2].

Very briefly maintenance work had been carried out on a condensate pump in the morning, but this work had not been completed. The pipe leading from the pump was not completely sealed. The team left a message in the control room at the end of the shift, reporting the work had not been finished and not to start the pump, but this message was lost. Later around 9:55pm, the pump was started up and oil / gas immediately started to discharge from the pipe. This ignited and an explosion occurred – damaging firewalls and intruding into the control room, making this unusable and it was abandoned. The platform was shutdown and its oil and gas wells sealed off – in principle making the facility safe. However, another condensate pipe was ruptured in the explosion and this caused a secondary pool fire – which was a survivable event and would have burnt itself out as its inventory depleted. However, this fire impinged on the high pressure 16" and 18" gas riser lines from Tartan and Claymore fields. Feeds to these lines were not shutoff during the emergency as staff did not have permission to stop production – in fact this might have had only a limited effect as it would have taken many hours to depressurize these lines. The Tartan riser line ruptured at 10:20pm and this released tons/sec of gas and the platform could not survive this event.

Piper's own firewater system had been turned to manual as there were diving activities planned and it is not clear of the firewater system ever actuated. The Tharos support vessel started spraying firewater at 10:30pm and another support vessel, the Sandhaven, launched its rescue craft. Some minutes later further rise pipes failed and a major explosion created further damage and killed 2 crew and 6 rescued Piper personnel on the rescue craft. Shortly after midnight most of the platform sunk into the sea.

1.3 The Cullen Inquiry

The UK Government launched a public inquiry, chaired by Lord Cullen. A very thorough investigation of the evidence followed and the report [3] made 106 recommendations in 5 main areas: regulatory systems, control of work, hydrocarbon release controls, emergency management, and workforce engagement. While all the recommendations were important, the most significant was a switch from Prescriptive regulations issued by the Dept of Energy towards a Performance based regulation. The UK HSE issued the Offshore Installations Safety Case Regulations in 1992 [4]. Many other recommendations dealing with technical and organizational maters have been implemented and today things are very different to what they were in 1988.

2. The Piper 25 Conference

Last year, coinciding with the 25th anniversary of the Piper Alpha event, the Oil & Gas UK sponsored a major conference in Aberdeen with the objective to: "to reflect, review, reinforce and re-energize". The Piper Alpha event and its lessons were summarized by several speakers and the points they made are directly relevant to US operations as well.

The Piper 25 Conference [5] was organized into parallel sessions. The sessions included:

- The main plenary event
- FABIG Session: Advances in Fire & Explosion Engineering
- Major Hazard Management and Safety Technology
- Emergency Preparedness and Response
- Safety Management Systems and Control of Work
- Human and Organizational Factors

Main plenary speakers included: Lord Cullen, Malcolm Webb, Judith Hackitt, Jane Cutler, Justice Haddon-Cave, Prof Andrew Hopkins, Bob Fryer, Jake Molloy, John Bresland, Thad Allen, and several others, plus the many presenters for the technical sessions.

3. Key Messages from Plenary Speakers

3.1 Lord Cullen – Chairman of Piper Alpha Inquiry

Cullen reflected on his Inquiry task. He recognized early on that regulatory updates would be needed, but that "no amount of detailed regulations could make up for deficiencies in the way that safety is managed by operators". Beyond the technical issues leading directly to the incident, the Inquiry addressed:

- Effective safety leadership
- Commitment to give priority to safety
- Safety representatives: their role and training

He reflected on three recent disasters: Buncefield, Texas City, and Macondo (two onshore and one offshore incident) and these show that the problem is not yet completely solved. He felt that more effort was required to fully extract the value presented by the safety case approach. However, he stands by the ideas behind goal-based regulations.

3.2 Hon Justice Haddon-Cave – Chairman Nimrod Inquiry

Justice Haddon-Cave chaired an inquiry into the crash of a Royal Air Force Nimrod in Afghanistan in 2006. The cause was purely technical, associated an air-to-air refueling after which fuel dripped onto hot pipes, ignited, and soon after brought the plane down, killing all on board. He highlighted 7 issues that developed over a period of 30 years.

Table 1. Sequence of Issues Leading to Nimrod Accident

Issue	Implication
Poor design and modifications from 1960's	Allow fuel to contact hot pipes
Fuel leak incidents observed in 1970's – 80's	No response results
Operational tempo increases – Kosovo, Iraq, Afghanistan	Greater stress on aging equipment
Maintenance delays due to operational demands	Extend service dates arbitrarily
Continuous organizational changes 2000-05	Strategic defense reviews leads to multiple change
Outsource Safety Case authorship to airframe builder	Much paper, not linked to real operations
Leadership / culture issues	Safety was not top priority

The inquiry brought out clearly several organizational influences, with 12 direct causation parallels to the NASA Columbia crash investigation. Some of the parallels included over-confidence in own skills, imposition of business practices over engineering principles, continuous organizational change, continuous cuts to resources, excessive outsourcing, normalization of deviance, and success engendered over-optimism.

Justice Haddon-Cave outlined 7 main lessons:

1) Focus beyond the immediate cause towards on organizational issues
2) Faulty assumptions – it has flown safely for 30 years – it must be ok; the safety documentation is extensive and says it is ok
3) Excessive organizational change is detrimental to safety
4) Avoid the comfort blankets of complexity, compliance and consensus.
5) Outsourcing is ok, but don't outsource the thinking or the skills to interpret the results
6) A Safety Case is an aid to thinking – not the end result for safety. The Nimrod safety case became a paper exercise with the aim of finding the aircraft was safe
7) Age matters – older equipment can be safe but it needs more rigorous attention and maintenance

He supports the idea of safety case, but it needs to be easy to understand and communicate. He likened it to new architecture – all the workings (pipes, electrical conduit) need to be on the outside and easily seen. As an interesting example of organizational complexity he showed the initial organization for Nimrod airworthiness compared to the final form.

Figure 1. Airworthiness organizational Structure

Initial organizational structure	Organizational structure after changes
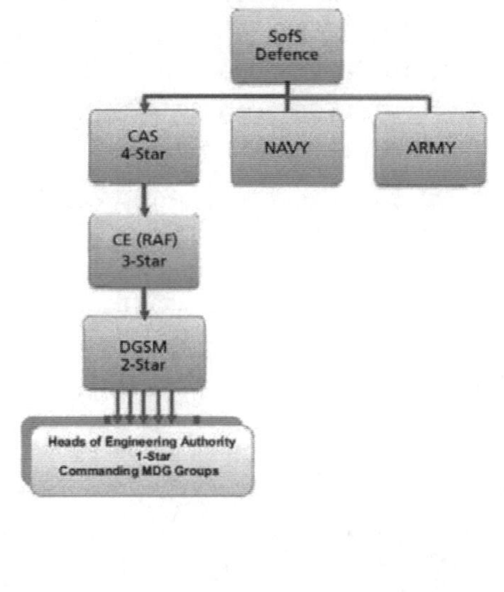	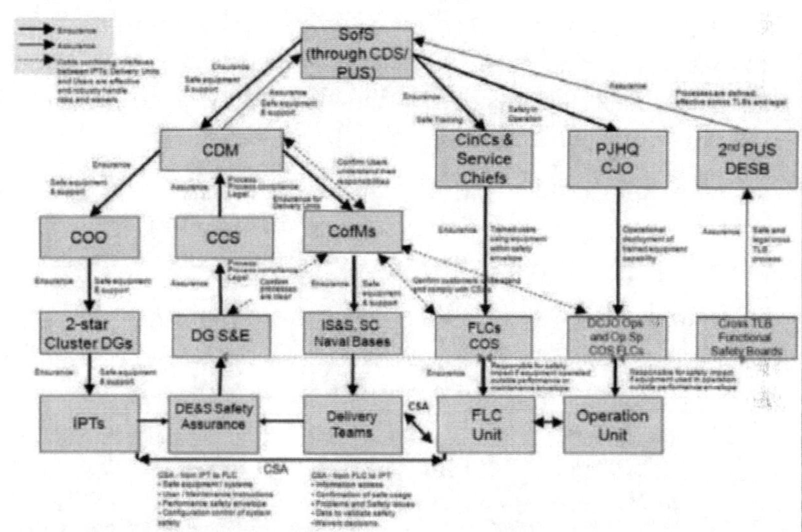

3.3 Judith Hackett – Head UK Health and Safety Executive

The UK HSE has tried to stay true to the recommendations contained in the Cullen Inquiry. They have implemented a performance based set of regulations for offshore:

- Goal setting regulations
- Operators are responsible for managing safety
- Operators must demonstrate their assessment of risks and measures taken to control them
- Support and encourage the workforce.

She highlighted the issues with a Prescriptive regulatory regime: Prescription can place the onus on the regulator to devise potentially complex solutions, and it can lock the industry into outdated solutions. Instead goal-setting places the onus on operators to devise solutions that are "As Low As reasonably Practicable".

She stressed the theme is learning. Use learning to shape what we do and improve, use learning to do things differently, and to be able to demonstrate that we have learned effectively.

She also had lessons for the regulator. Do they have the right resources and are they making best use of those, are they making the right interventions to achieve extra safety, and do they properly monitor and review their own activities.

She ended with a quote from Charles Darwin: "It is not the strongest of the species that survives, nor the most intelligent, but the one most responsive to change."

3.4 Jane Cutler – Head Australia NOPSEMA (Offshore Regulator)

Australia took on board the Cullen recommendations to establish goal setting regulations underpinned by safety case by 1996 – without having a major offshore accident to drive this. The National Offshore Petroleum Safety Authority (NOPSA) was established in 2005. This has issued many guidance notes and obtained industry participation – although with some weaknesses. This was exhibited in 2 major accidents – a 12" pipeline rupture on Varanus Island (West Australia, 2008) and the Montara blowout (Timor Sea off Northern Territory, 2009). Some weaknesses were: regarding the safety case as a compliance document – it needs to be written and approved; regulator assistance needed too often for successful completion – by means of submission and rejection with advice on to how to fix; and the Safety Case even where well written is ignored in daily operations – as found in the Montara Inquiry. All of this leads to safety case as a paper exercise defeating its objective of stimulating thinking, properly managing risks, and communicating these to the workforce.

The Montara Inquiry led to further focus towards a single competent regulator – NOPSA authority was extended to cover environmental impacts (now NOPSEMA), and it became responsible for drilling – previously with the States.

She commented on issues for the regulator and these were similar to the HSE comments. Do we have the right resources? Are we making best use of skill sets in our organizations? Is there the right mix of interventions to affect sustainable change? She concluded with some challenges remaining since the time of Piper Alpha and still not resolved:

- No comprehensive international incident database existed then; none exists now
- No comprehensive international database for safety alerts existed then; none exists now
- Standards development was disjointed then and remains so today.

3.5 Bob Fryer – Executive VP Safety and Operational Risk, BP

Mr Fryer made many observations on changes since Piper Alpha and more recently since the Texas City and Macondo events. The industry has made major improvements to occupational safety, but rather less so for major accidents. In BP lost time injuries have fallen from around 5.7 LTI/million hours at the time of Piper Alpha to 0.43 in 2012. The UK Step Change in Safety initiative was a major impetus. More recently Macondo has been a driver for change within BP. The BP Bly investigation report made 26 recommendations and 14 have been completed so far. An independent safety adviser has been appointed by the Board. Offshore specialists have been recruited and BP now has in-house 12 cementing and 30 BOP specialists. For general skills enhancement BP has established a Global Wells Institute and a wells simulator (located in Houston) for use by BP staff and contractors. There is also a well monitoring center where staff offshore can communicate easily with specialists onshore.

BP has changed organizationally. It moved from an asset model – with a regional organization, to a functional model – with a global structure – so exploration staff report to a global head of exploration, projects staff report to a global head of projects, etc. This provides for stronger technical management. BP have established a Safety and Operational Risk group with real authority to control changes deviating from normal operations. It has responsibility for:

- Setting safety and operational risk requirements
- Carrying out assurance and audits to maintain an independent view on risks
- Maintaining deep technical expertise to advise or intervene as needed

BP has created a company-wide risk management systematic – the Operating Management System which uses amongst other tools the swiss cheese / bow tie barrier model. This defines critical barriers and accountabilities. There is a risk assessment program being implemented in all businesses followed by risk action plan reviews – to address the identified risks.

Finally BP is committed to recording and reporting process safety events (based on the API 754 system). Initial results are available and 2012 saw a 42% decline in Tier 1 loss of containment events compared to 2011.

3.6 Steve Walker (UK HSE)

Steve Walker (HSE) reviewed the ongoing relevance of the Cullen recommendations. The UK offshore sector is not stagnant – there is aging infrastructure, new global players, with deeper and more challenging reservoirs, coupled with public concerns over environmental spills and a desire for more open sharing of incident data. He noted that Prof Maitland's review of the UK regulatory approach for offshore in 2011 after Macondo. He fully assessed its value and he had commended it – particularly aspects relating to goal-setting and independent 3rd party verification of well design and safety equipment. Also safety case has been adopted in Australia and the new EU Offshore safety regulations – so this is 3 independent validations of the goal-based safety case approach. The HSE finds that control of work remains a persistent problem. Electronic PTW systems have improved the process compared to Piper Alpha time – and should have prevented that accident, but there remains a lack of compliance to permit requirements and incorrect isolation. The Cullen recommendations on reducing hydrocarbon leaks have been achieved as well as full reporting of leaks. He noted the decline in major leak events with an ongoing commitment to reduce these further. Emergency response systems have improved considerably since the time of Cullen (equipment on rigs and the BP Jigsaw helicopter system), but the HSE still finds issues with TSR's (Temporary Safe Refuges) and effectiveness of damper closures. Workforce engagement has improved greatly since the time of Piper, and there are many effective programs currently (award programs, engagement toolkit, etc.). However recent HSE inspections in 2011 have revealed variable implementations offshore. Overall, he concludes that the Cullen recommendations remain revelant today.

3.7 Other Plenary Session Comments

Some other key comments made during the plenary session included:

Andrew Hopkins strongly supported the Safety Case concept combined with the duty of care, and that it needs to incorporate organizational factors in its thinking.

John Bresland highlighted learnings from the US Chemical Safety Board by reference to several accidents. Key needs were: to demonstrate leadership in process safety, that excellence in process safety is a key to your success, to take the long view with risk, measure both personal safety and process safety, and to avoid complacency.

4. Technical Sessions

4.1 FABIG Session (Fire and Blast Interest Group)

This session covered several technical aspects used for offshore design and how these have changed since Piper Alpha.

In the fire domain, Geoff Chamberlain (ex-Shell Thornton Research Center) showed the many improvements in fire knowledge since 1988. For pool fires modeling parameters were only well established for open fires. Other fires such as fuel or ventilation controlled fires in modules were poorly known. Fires on sea surface were roughly known and fires originating with spills below the surface were hardly understood. By 2013 – the situation on all of those had improved so that good modeling was possible for all types of pool fire.

A similar situation exists for jet fires. In 1988, open gas jet fires were well understood, but major uncertainties were associated with flashing liquid jet fires, confined fuel or ventilation controlled jet fires, and jet fires on sea surface. By 2013 – again better knowledge was available for all these jet fire types. Fire modeling has improved from solid flame models predicting thermal loads to CFD models predicting temperature response in structures. There of course remain unknowns. Free radical fire reaction kinetics and modeling is still advanced research, although Sandia used this method for their recent LNG pool fire predictions [6].

Kees van Wingerden (Gexcon) presented the change in explosion knowledge since Piper Alpha. At that time explosion approaches were rather prescriptive with poor underlying assumptions, venting was by correlations and ignoring obstructions, and correlation models were most commonly used for explosion prediction (e.g. the TNO Multi-Energy Model) – but these were developed as onshore far field models.

The response after Piper Alpha was to carry out significant experimental work to fill the knowledge gaps: HSE / Industry Phase 1, 2, 3a and 3b experiments, the EU sponsored projects MERGE & EMERGE, FABIG ongoing work, and in the USA the Explosion Cooperative BakerRisk JIP. Simultaneously with the experiments, model tools also developed, particularly the CFD tools (e.g. FLACS). It is now routine to carry out CFD analysis of design alternatives to establish necessary separations, structural strength requirements or potentially blast wall mitigations.

This paper also showed what extra knowledge would have been available to Occidental if modern CFD tools had been used on the Piper Alpha design.

Other papers in this session addressed changes in structural protection (Ramsay Fraser, Atkins) and the use of advanced safety models for offshore facility design (Asmund Huser, DNV GL).

4.2 Major Hazard Management Session

Doreen Gray (Scottish Power) shared an integrated system of safety performance indicators that they have developed and deployed across their sites. This is based on an underpinning bow tie model of risk identifying 42 risk control systems and over 100 leading indicators (Figure 2). These are converted on a daily basis into easy to understand metrics and pushed to all staff through a common interface. The system has previously been noted to be amongst the most sophisticated major accident KPI system currently deployed.

Figure 2. Scottish Power KPI System

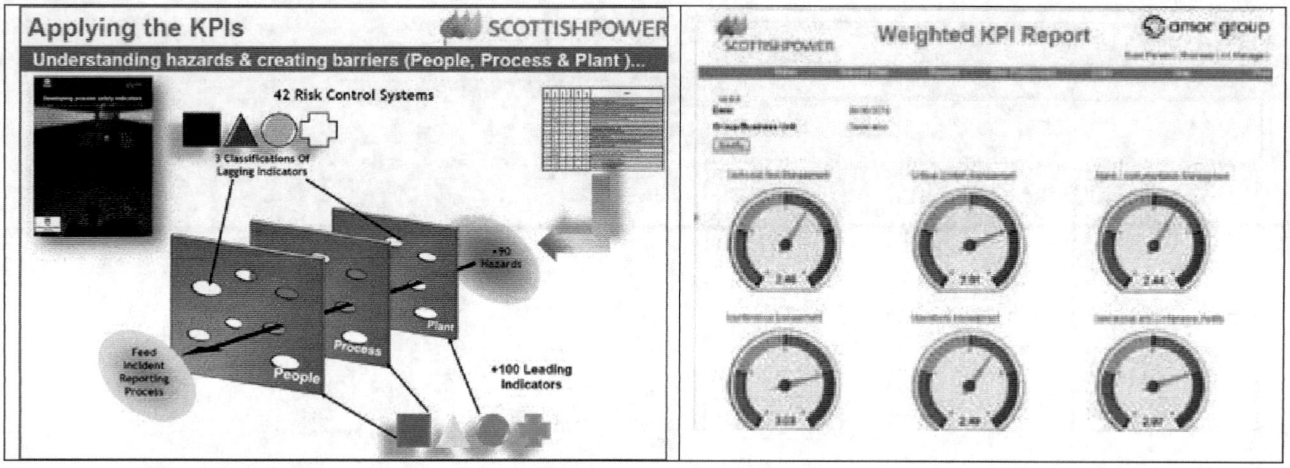

Nikki Jackson (BG) showed a barrier based Cumulative Risk Management (CRM) Strategy being implemented by BG. This starts with the basic risk but adjusts this for equipment out of service or beyond inspection intervals. The updated risk is then used as the basis for decision making (Figure 3).

Figure 3. BG Cumulative Risk Management System

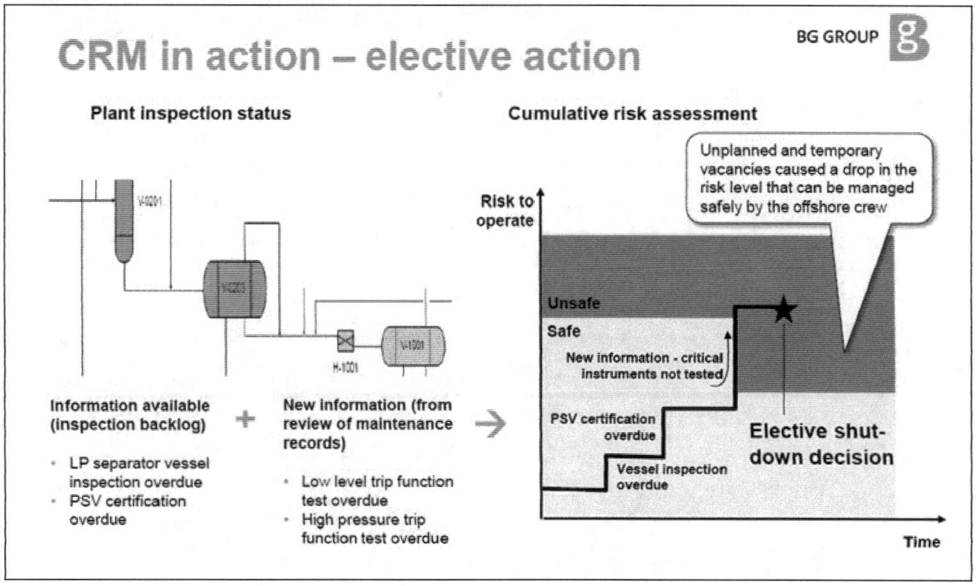

James Phipps (Petrofac) showed how they have been using the full features available from advanced Bow Tie drafting tools (e.g. BowTieXP) to manage major hazard risks. Few companies use the full range of tools available and this example showed the potential to good effect. As well as the basic bow tie diagram with color coding for accountable people or barrier features; barriers can be linked to safety critical elements, audit findings, and key documents supporting the barrier performance (Figure 4). Some more advanced uses now under development include an ability to establish the current live barrier status – fully functional, degraded, or broken.

Figure 4. Advanced Bow Tie modeling features

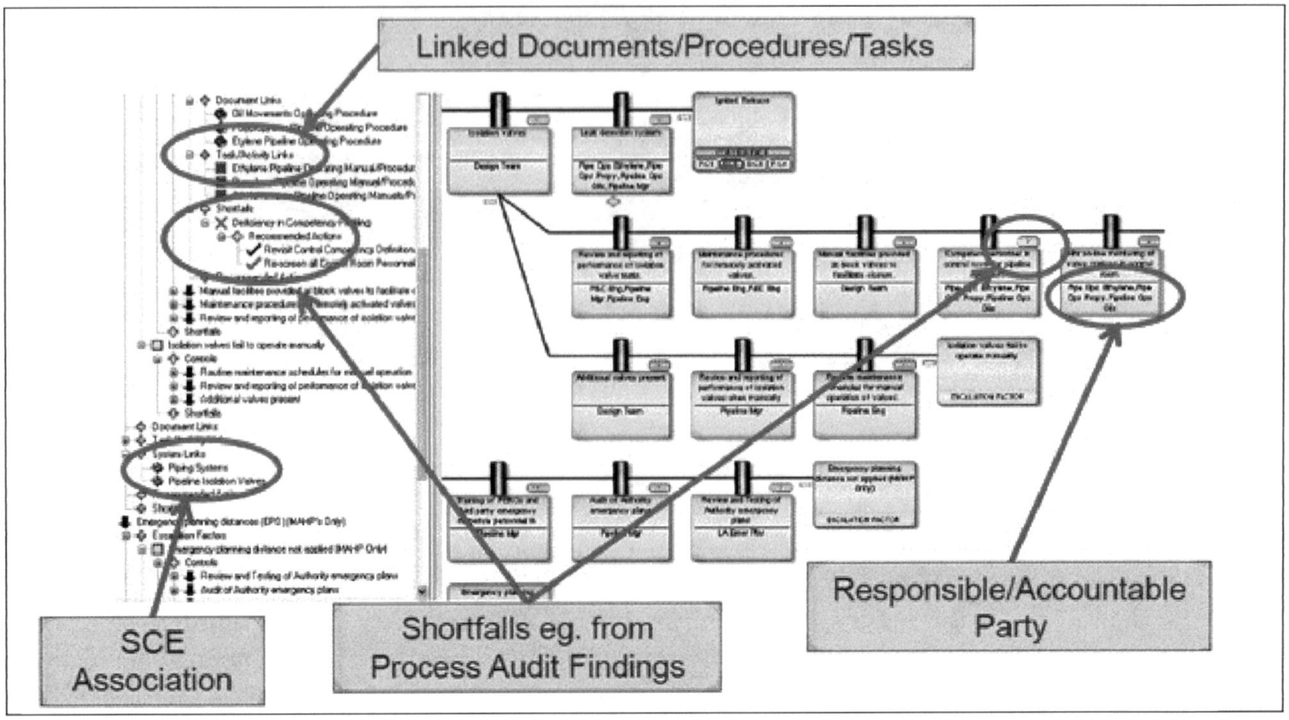

Peter Boyle (DNV GL Aberdeen) reflected on changes since Piper Alpha. While there have been many positive improvements, as listed by others, there remain many tasks still to address. Whilst occupational safety has improved by about one order of magnitude in the time since Piper Alpha, there has hardly been any improvement on reducing major accidents both offshore and onshore is a challenge to the O&G industry (Figure 5). DNV [7] issued a position paper that makes a number of suggestions for enhancement – many but not all reflected in other papers at this conference. These key areas include: performance based regulations, clear roles and responsibilities, holistic risk approach, shared performance monitoring, advanced barrier management, and focus on people – plant – process.

4.3 Emergency Preparedness and Response

Pete Thompson (EPOL Group) surveyed the difference in emergency response available since Piper Alpha. At the time the Piper Alpha response was the largest UK response to a major incident offshore (and remains so today). A total of 53 surface vessels, 19 helicopters and 4 fixed wing aircraft responded – this is dwarfed by the scale of response to Macondo. Many operational difficulties were experienced in the coordination of the response. Many enhancements have occurred since then. These are a mixture of organizational – Emergency Preparedness Offshore Liaison (EPOL) Group established to coordinate amongst all key responders, regulatory – PFEER Regulations were passed Prevention of Fire Explosion and Emergency Response, a new Offshore Emergency Response Working group was established after several accidents to review and enhance emergency response and search and rescue activities. Note this paper describes the human safety response aspects and is different to the OGP post Macondo oil spill response facilities.

Figure 5. Timeline of major offshore incidents

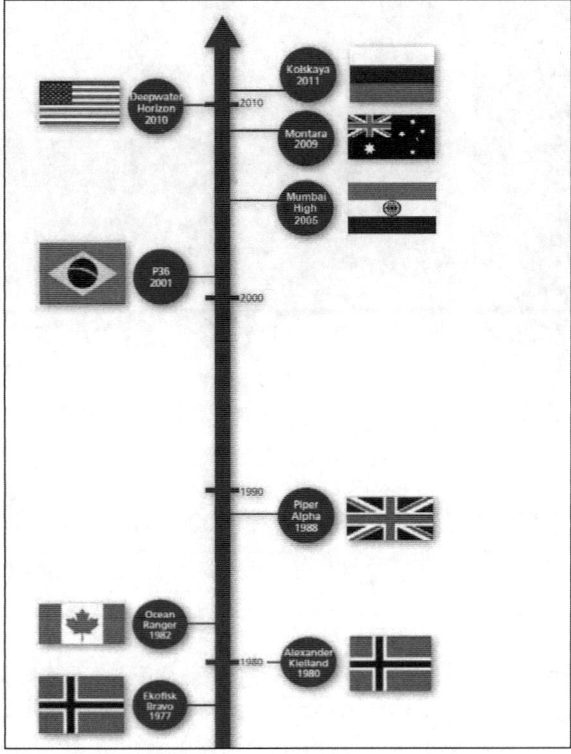

This timeline of offshore major accidents shows only some more serious events – the time starts before Piper Alpha and concludes in 2011.

A rather obvious message is the multiple countries represented and the lack of any decline in the trend.

4.4 Human and Organizational Factors

John Wilkenson (Keil Center, and past HSE Human Factors Lead) presented his impressions of change. In 1988 human factors was hardly discussed. Now, it is included in the safety case, in safety critical tasks, included in HOF (human and organizational factors) investigations, Just Culture, part of High Reliability Organizations implementations, etc. Some time ago the HSE identified 10 main areas for human factors enhancements – while there has been progress, many issues remain with more to do.

These 10 topics were: managing human failure, procedures, training and competence, staffing, organizational change, safety critical communications, human factors in design, fatigue and shiftwork, organizational culture, and maintenance inspection and testing.

Gary Begg (Talisman Sinopec) highlighted activities carried out by his company addressing human and organizational factors after an uptick in their reported leak frequency. He described their approach as "joined up thinking" and this did result in getting back onto the downwards leak trend. Key factors in success were strong leadership & workforce engagement. These success elements match closely many other comments made – but of course leadership and engagement are only effective if you have the right program. Talisman Sinopec drew down on excellent documents produced by the UK Step Change Initiative [8]. These are practical documents well suited to workforce engagement.

Figure 6. UK Step Change Publications

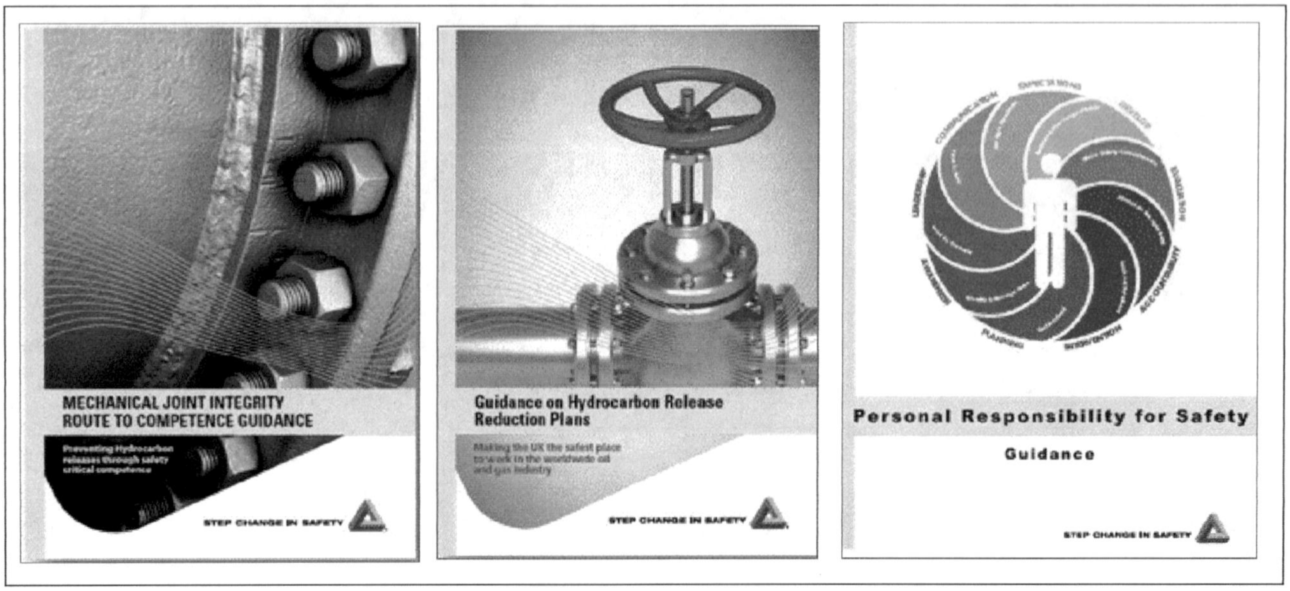

5. Conclusions

The Piper Alpha event was a major wake-up call for the industry – too often good intentions and good ideas can decline over time. Many positive actions have resulted from all key stakeholders. The UK switched to performance based regulations, industry worked hard on implementation, design standards improved and the science underpinning fire and explosions was greatly extended. Key areas have shown improvement including: process safety technology, management systems, barrier management, emergency preparedness, and human factors. Although perhaps that last topic has the greatest challenges remaining. But as one commenter said, companies all now recognize the importance of human factors for safety and it is not just dealing with "fluffy stuff". The presentation by Justice Haddon-Cave on the Nimrod air crash – although not from the Oil & Gas industry was widely regarded as thought provoking and offering lessons of great importance to all industries handling aging infrastructure involving high technology. The overall conclusion was that the Cullen recommendations offer as much value now as then, and ongoing efforts remain to address some weaknesses identified by the HSE.

6. References

[1] Lees F. P., "Loss Prevention in the Process Industries", 4th Ed. (3 volumes), Edited Sam Mannan, Butterworth-Heinemann, 2012.

[2] Atherton J. and Gil F., *Incidents that Define Process Safety.* Center for Chemical Process Safety, New York, Wiley-AIChE, 2004.

[3] The Hon Lord Cullen, Public Inquiry into the Piper Alpha Disaster, UK Dept of Energy, 2 Vols, 1990

[4] UK Offshore Installations (Safety Case) Regulations 1992 (SI 1992/2885) HMSO Publications.

[5] Piper 25 Conference, Oil & Gas UK, Aberdeen, June 18-20, http://www.oilandgasuk.co.uk/events/P25Agenda.cfm,

[6] Liquefied Natural Gas Safety Research, Report to Congress, Sandia National Laboratory, US Dept of Energy, Sandia report to Congress, May 2012.

[7] DNVGL, Enhancing Offshore Safety and Environmental Performance, Position paper, available from DNVGL Oslo or Houston, 2013. http://www.dnv.com/industry/oil_gas/offshore_safety_position_paper.asp

[8] UK Step Change in Safety, cooperative group established in UK, publications available: http://www.stepchangeinsafety.net/

BP's Process Safety Journey: Enhancing Risk Management in Global Upstream Operations

Dr Steven A Flynn, BP plc

Summary

BP has been implementing a global effort to enhance safety and operational risk management in operations and projects, which has incorporated the learning from the Deepwater Horizon and other high profile industry incidents. This paper provides a brief update on the status of drilling safety and oil spill response improvements, and then focuses on the wider process safety-related enhancements aimed at improving risk management, with examples from BP's upstream production operations. The overarching strategy is built of three enduring principles:

- People - deep capability and a safe operating culture
- BP's Operating Management System (OMS) is the way BP operates
- Self and independent assurance confirming conduct of operating

As part of the OMS, a single BP-wide framework is used for assessing and managing operating risks. In the Upstream business, this has prompted a number of global risk reduction projects, including focus on:

- Pipeline corrosion
- Control systems, alarms and safety instrumented systems (CASIS)
- Relief systems

These are part of on-going efforts to strengthen engineered and procedural barriers and to learn from observed trends. As these programmes have progressed, both process safety metrics and plant reliability have been observed to improve.

1. Introduction

The learning from Deepwater Horizon and other high profile incidents within the industry have been taken forward through global efforts, both within BP and across the industry, to enhance safety and emergency preparedness. This has led to enhancements in drilling safety and upgraded oil spill response capabilities. BP has also built on this to further enhance safety and operational risk management across all its global operations activities, including upstream, downstream and alternative energy.

2. Drilling Safety and Oil Spill Preparedness Update

Progress on drilling safety and enhanced oil spill response capabilities has been wide-ranging, and regularly shared through joint industry programmes, technical conferences and various publicly available reports. For example, more than 2,000 people have been brought together into a single BP Global Wells Organization. Enhanced drilling standards have been implemented for operations and equipment, including blowout preventers, cementing and oil spill response capabilities, and oversight of contractors. A programme to enhance assurance of drilling competency for key positions is also underway.

BP, together with the industry, has continued to strengthen capability to respond in the event of a loss of well control. For example, a global deepwater capping and containment package comprising 250 pieces of specialty equipment is in place that, under ideal circumstances, can be flown to any country where BP operates within a few days. BP is also advancing oil spill response processes and technologies, and participating in various joint industry projects.

The challenge to embed and apply learning has also resulted in progress across the world-wide industry. For example, the Oil & Gas Producers Association's (OGP) Global Industry Response Group is progressing in three main areas:

- The Wells Expert Committee is well analysing incident data, advocating harmonised standards and good practice, and promoting research and development
- The Subsea Well Response Project has placed well capping and dispersant application equipment for use by the industry in four regional bases: Rio de Janeiro, Stavanger in the North Sea, Saldanha in South Africa, and Singapore.
- Oil Spill Response Joint Industry Project is developing new guidance and recommended practices, and enhancing co-ordination between key stakeholders internationally.

This is just a small sample of continued enhancements at the company, region and global industry level.

3. Company Wide Focus on Safety and Operational Risk

The global safety and operational risk program within BP covers the breadth of BP's operations. This includes upstream exploration, development and operations; downstream refining, chemical and product manufacturing, and marketing activities; and other business activities such as shipping, wind power and biofuels. The program is built on the fundamental belief that systematic operating is the foundation of a safe and strong company. It drives both safety and efficiency, and therefore long-term value. Said another way: 'A safe business is a successful business'.

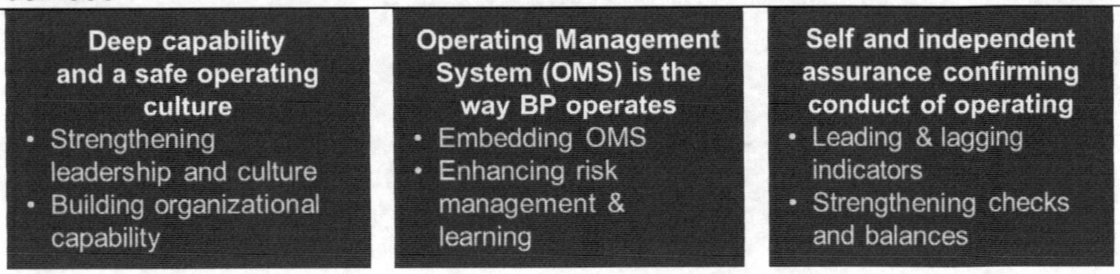

Fig 1. Company Wide Changes on Safety and Operational Risk – enduring themes

The programme of action is aligned with three enduring principles which together provide a framework for safe, reliable and efficient operations (Fig 1). Under these higher level themes, there are specific areas of focus that drive group-wide work programmes:

- People - deep capability and a safe operating culture
- Operating Management System is the way BP operates
- Self and independent assurance confirming conduct of operating

The following sections provide an overview of each of these areas as context for an in-depth review of risk management.

3.1 Deep Capability and a Safe Operating Culture

The first enduring theme relates to people – both deepening capabilities, and a strong emphasis on safety leadership. The tone is set from the top with the CEO setting clear expectations of the company with respect to safety: that safety lies at the heart of BP and the foundation of good business. Leaders are expected to spend time in the field, engaging with staff on the front line to understand operational risks. Tools and guidance on how to support effective use of this time in the field are being provided, focusing on risk management barriers in their operations.

Safety and operational risk education programs have been enhanced to better equip our leaders and operators with the skills they need to perform and lead systematic operations. These programs reach from senior operations leaders down to the front line. One example is the Operating Essentials program for front line staff that is delivered on site among teams. More than 6000 people typically attend during a year. This program focuses on skills in areas such as Continuous Improvement techniques, Control of Work and Hazard Identification.

3.2 Operating Management System (OMS) is the Way BP Operates

The Operating Management System or 'OMS' is the foundation of the way we work in BP operations. It is used to drive systematic operating and to manage safety and operational risk. It is a framework that sets out 'what' has to be done and how', and it has an annual cycle designed to drive continual improvement (Fig 2).

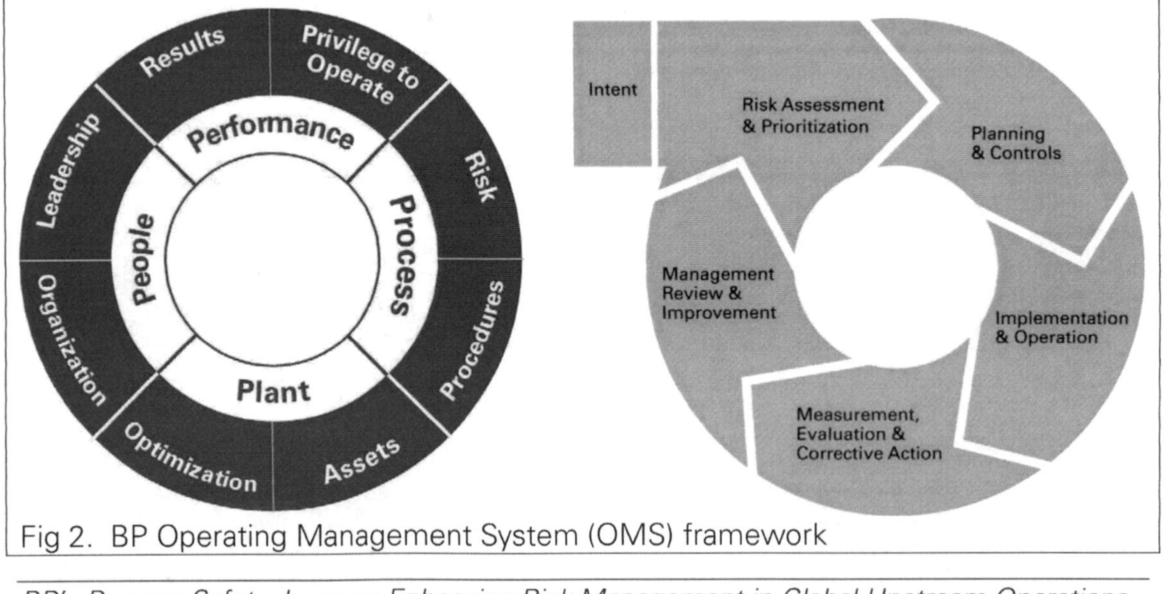

Fig 2. BP Operating Management System (OMS) framework

The 'what' part of OMS (shown on the left) has 8 elements covering People, Plant, management Processes and Performance. Where necessary, the high level requirements are backed up by company requirements, procedures and guidance. These include group wide requirements for safety and operational risk assessment and management, incident investigation and learning, and specific process safety practices such as HAZOP, Layer of Protection Analysis (LOPA) and Safety Instrumented Systems (SIS).

The 'how' part of OMS is the Performance Improvement Cycle (shown on the right). It is designed to run at least annually to identify, prioritize and implement planned improvements. The OMS is fully operational and we have also created ways for sites to learn from each other, including a program called 'Exemplar' which brings specialist coaches onto sites to help them accelerate improvement in particular areas.

3.3 Self and Independent Assurance confirming Conduct of Operating

The third principle is 'assurance' – essentially assuring ourselves as to whether relevant capabilities are present and OMS is working to continuously strengthen our operations. A three-tier approach to assurance is used (Fig 3). Monitoring performance metrics is one way of doing this and BP's approach to leading and lagging indicators is aligned with API RP 754.

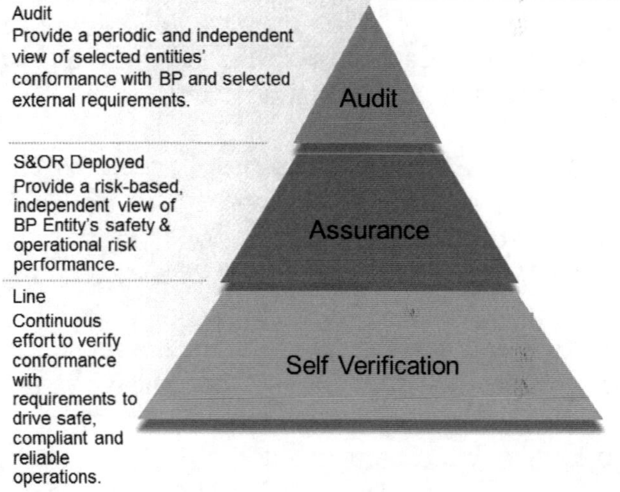

Fig 3. The 'Three Lines of Defence' in assurance.

- Operating line management are accountable for safety, they conduct self-verification to confirm whether they are conforming to OMS and that risk barriers are robust, and to enable them to take action as needed. This is the first line of defence.

- Deployed-staff from the independent Safety and Operational Risk Function, conduct targeted, risk-based assurance by regularly checking to see how operations are conforming to requirements and managing risks. This second line of defence operated in a structured company-wide way can be used to determine if there are risks to be addressed more broadly.

- Independent audit is the third line of defence. A dedicated team within the company's group audit function conducts a risk-based programme of regular safety and operations audits. This includes auditing of third party rigs and ships for conformance with BP's requirements.

For example in 2012, thirty-four contractors' rigs were audited, and actions identified to address the findings. This included thirteen new rigs – with a number of actions identified to be addressed before they started work for us. One rig was held back for three months while the relevant actions were addressed and we have decided not to use rigs that do not meet requirements.

4. Company Wide Safety and Operational Risk Management Process

The many procedures used to manage risks on a daily basis integrate into a single BP-wide framework which is used for identifying, understanding, managing, reducing, and if possible eliminating, operating risks. While the focus remains on day-by-day risk management, every BP operation performs an annual "snapshot" review of operating risks, updated as necessary if there are substantial changes. This also provides an input to our annual business planning process and enables resource allocation to be focused where it is needed most. During this annual review, the operations confirm that controls are in place and set priorities for further risk reduction or elimination.

In the 2013 risk process, more than 50 businesses, comprising many hundreds of facilities, assessed risks and conducted risk action plan reviews which were the inputs to 2014 business plans. The process engages the company at all levels – from frontline operations to the group executive and Board.

Data is fed into this process from specific process safety processes. For example, the Major Accident Risk process is a coarse form of quantiatative risk assessment that is used, where appropriate, to assess potential major accident hazards and prioritise possible reduction projects. A software tool has been developed for offshore facilities, allowing a relatively quick assessment of risk considering the three dimensional layout. Other risk management tools include Inherently Safer Design, HAZOP, Layer of Protection Analysis (LOPA) and Safety Instrumented Systems (SIS), . Application of these processes have led to tangible risk reductions through risk elimination, reduction and control.

As part of this risk reduction process there have been tangible changes to operations including the following for upstream:

- Control rooms and office buildings on operating sites have been upgraded or relocated. For example, a brick constructed control room on an upstream production centre was replaced with a more blast compliant structure. At another gas processing plant, non-essential office staff were relocated offsite.

- Processes were modified for safer alternatives. For example, ammonia-based refrigerants were replaced with less toxic alternatives at a gas processing plant.
- Control systems have been upgraded and a project for assuring safety instrument systems is being implemented across upstream operations.
- Inherently safer designs have been employed, particularly in offshore developments. For example, not normally manned facilities and designs that minimize visit frequency are used for some offshore production platforms. The Clair Ridge project in the North Sea has been designed with a bridge linked accommodation platform, which significantly reduces the risk to personnel and layouts on floating installations, layouts have been optimized to minimize potential fire and explosion impacts.

5. Upstream Risk Reduction – Priority Projects

In addition to identifying individual risk reduction projects, reviews of risk assessments, incident trend analysis, and findings from audits and assurance activity have identified broader risk reduction programmes for application across multiple upstream operations.

One of the tools proving particularly effective for communicating and managing risks is the 'Bow-Tie' model, which is well known in the process industries (Fig 4). A bowtie depicts the barriers put in place to prevent and mitigate a potential event. This supports a deep and consistent understanding of the risks and can be used to help communicate to the front line and management alike in further driving down risk.

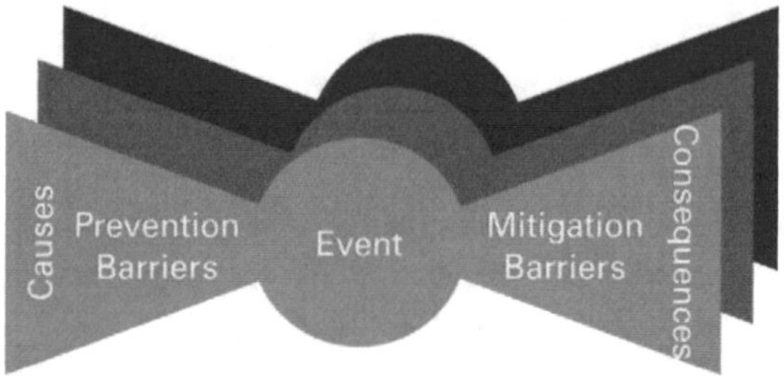

Fig 4. Risk 'Bow-Tie' principles.

For example, a Bow-Tie model can be developed for loss of primary containment of hydrocarbons that could result in safety and environmental consequences such as fires, explosions or spills (Fig 5). The key initiating events in the bowtie below are:

- Operating outside of the safe design limits, due to control system issues
- Degradation of equipment, such as mechanical integrity of production systems or pipelines
- External factors, including marine collisions with offshore installations and cyber threats

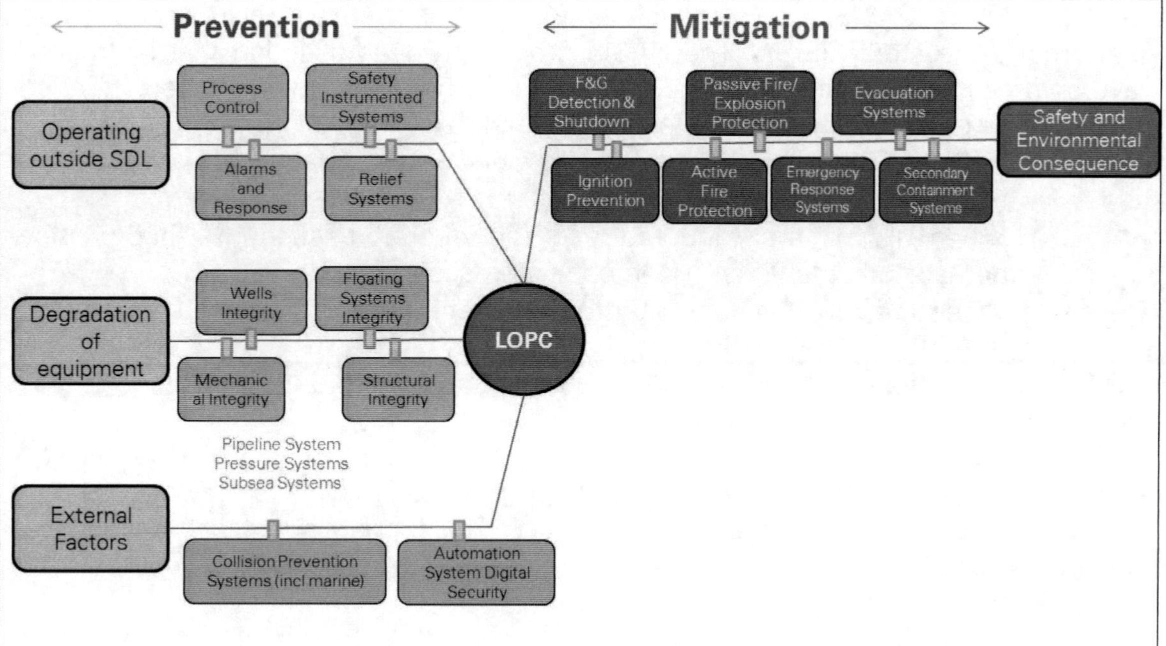

Fig 5. Generic Bow-Tie Model for loss of primary containment (LOPC) in upstream operations.

While systems and OMS requirements are in place to manage these threats, focused assurance of barrier strength can help further reduce risks, identify potential gaps and spread learning across the portfolio. This approach can be illustrated by three global priority projects to further enhance management of plant integrity and process safety in our upstream operations (Fig 6).

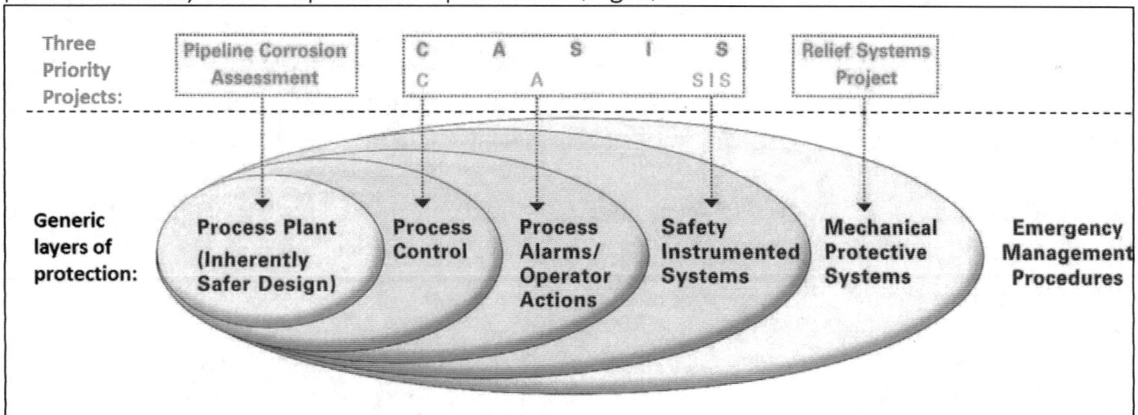

Fig 6. Upstream priority risk reduction projects.

The first of these is pipeline corrosion assessment. Corrosion can occur in the oil and gas industry due to the fluids handled and the environmental conditions. Assessing and predicting corrosion is a part of our integrity management strategy. We have developed a new, standardized methodology to assess the potential corrosion threats, prescribe effective barriers, and provide ways for verifying that those barriers are doing their job.

The second is 'CASIS' – which stands for control systems, alarms and safety instrumented systems (SIS). The CASIS project involves assuring that instrument layers of protection are in place, and that they will work when needed. Strengthening one layer reduces the demand on the next.

- By keeping the process within operating limits, the demand on alarms is reduced.
- When alarms are triggered and the operator responds appropriately, demand on safety instrumented systems is reduced.
- When the SIS functions as intended, demand on the pressure relief and flare system is reduced.

All of these components provide layers of protection to help deliver safe and reliable upstream operations.

Guidance has also been provided on topics such as SIS lifecycle, safety override and risk assessment (SORA), and alarm rationalization to support operations, testing and maintenance.

The third project is relief systems - the final line of defence. The intent is to verify that the pressure relief and flare systems will perform as intended is called upon.

These projects are one aspect of an enhanced focus on self-verification in upstream operations. Another example is the focus areas that have been identified by global operations leadership for self-verification in day-to-day production operations, which include items such as permits to work and plant isolations.

6. Learning and Improving

Complementary to the forward looking risk identification and management processes and the enhanced layers of assurance, there is a global programme to further improve learning from incidents and near misses. Structured mandatory methods for investigating, learning and taking corrective actions are in place as part of the OMS. Specific steps to improve the quality of investigations and learning in our world-wide operations include:

- Bringing in world-class expertise, including former US FAA and NASA experts, to bring learning from other industries
- Enhancing investigation and learning processes and tools, and the capability of the investigators
- Independent input to the execution and review of key investigations by independent S&OR experts
- Executive leadership review of top category incidents and near misses.

The company-wide approach to embedding learning from incidents is formalized in our High Value Learning (HVL) process, which requires positive confirmation from operating sites that relevant actions have been completed.

7. Final Thoughts – Process Safety Performance

This paper focuses on some of the specific actions being taken within the wider process safety improvement programme across the BP group of companies. Some process safety metrics are included to provide a picture of progress. These figures have not been adjusted to reflect BP's recent divestment programme but the same underlying trend in the company's process safety performance can be seen with or without the divestments.

At the group level there have been continuing downward trends in process safety events and losses of primary containment (LOPCs) over several years.

For BP globally, Tier 1 PSEs reduced 73% from 74 in 2010 to 20 in 2013 and a 48% reduction in the lower consequence Tier 2s. In Upstream Tier 1 PSEs we saw a 80% reduction in the same time period from 41 to 8.

BP first started reporting LOPC as a Group metric in 2008, in 2010 the number of LOPCs was 418 and in 2013 this figure has fallen to 261. And the leaks are smaller. In Upstream, there was a 26% reduction in LOPC to 143 in 2013 from 194 in 2010.

Of course, while some of the process safety data is encouraging, it is recognized that these metrics should be tracked over years. Analyzing these metrics is only part of BP's efforts to drive continuous improvement. It is recognized that any process safety event or loss of primary containment can have serious consequences.

And so the journey continues. Safe and reliable operations remain the top priority for BP and an enduring value for everyone. This starts with leadership and the tone, message and expectations from the top are clear. There has been a global effort to become ever more systematic in how we operate and our management system and risk processes are key components to drive continuous improvement. Process safety has been, and continues to be, a clear focus and our organisation's understanding of individual risks and barriers has been enhanced as a result. And while these efforts are making a tangible difference, we also know there is always more to do at BP and in the industry.

Development and Application of Spacing Chart for Simultaneous Operations (SIMOPs) in Unconventional Shale Gas Fields

Chandra Gulati
Shell Exploration & Production Co.
150 N Dairy Ashford, Houston, TX, 77079, USA
Chandra.gulati@shell.com

Julie Mialaret
Shell Exploration & Production Co.
Julie.mialaret@shell.com

Peng Lian
Shell Exploration & Production Co.
P.lian@shell.com

Duncan Smith
Risktec Solutions, Inc.
1110 NASA Pkwy., Suite 203, Houston, TX, 77058, USA
Duncan.smith@risktec.com

Prepared for Presentation at
American Institute of Chemical Engineers
2014 Spring Meeting
10th Global Congress on Process Safety
New Orleans, LA
March 30 – April 2, 2014

UNPUBLISHED

Development and Application of Spacing Chart for Simultaneous Operations (SIMOPs) in Unconventional Shale Gas Fields

Chandra Gulati
Shell Exploration & Production Co.
150 N Dairy Ashford, Houston, TX, 77079, USA
Chandra.gulati@shell.com

Julie Mialaret, Peng Lian
Shell Exploration & Production Co.

Duncan Smith
Risktec Solutions, Inc.

Keywords: enter a few keywords here to aid searching

Abstract

There has been a boom in development of North America's unconventional shale gas in recent years. Due to heavy reliance on equipment-intensive operations such as hydraulic fracturing, and an increasing desire to minimize surface disturbance and footprint, well site equipment spacing and layout is becoming a more complex issue for unconventional gas projects, particularly when there are simultaneous operations (SIMOPs) at the same well pad location. The Technical Safety Engineering team in Shell Exploration & Production Company developed a spacing chart for SIMOPS to help field supervisors and personnel make decisions about equipment layout for SIMOPs including drilling, completion, flowback, production, and construction.

Development of the chart employed a method using consequence modeling, which studies the physical effects of flammable gas dispersion upon release, and heat radiation from jet fire or pool fire. Release scenarios were identified for SIMOPs equipment from which release and ignition of flammable hydrocarbon may occur. Scenarios and minimum spacing distances were selected according to the principle of Managing Risk in Shell's [1] HSSE & SP Control Framework. The development process provides an opportunity to check through onsite hydrocarbon release and fire hazards in a systematic way for various well site operations.

The chart is used at Shell's unconventional shale gas facilities in the United States and is integrated into Shell's overall risk management system for SIMOPs. The aim is to use the spacing chart as a practical tool to manage the risk of fire hazards at well sites. Engagement with leadership of the various parties responsible for well site operations is necessary for successful chart usage in the field and establishment of a feedback channel for further improvements to the chart. Delivery of the chart to the field also enhances awareness of fire risks through communication and onsite training.

[1] Each Shell Oil Company subsidiary or affiliated company in the United States is a separate legal entity. In this document the term "Shell" refers to the subsidiary or affiliated company of Shell Oil Company that owns, operates, or provides services with respect to the assets and operations discussed.

1. Introduction and Background

Along with the recent boom in North America's shale gas development, Shell has expanded its onshore unconventional gas portfolio in the United States. Tight shale gas development operations involve different procedures and technologies than traditional onshore well operations. In particular, tight shale gas development depends heavily on hydraulic fracturing, which is an equipment-intensive procedure (Figure 1). In addition to increased complexity in field operations, more Simultaneous Operations (SIMOPs) are being carried out in the field. SIMOPs are scenarios in which two or more operations occur at the same time on the same well pad facility. SIMOPs include drilling, completion, workover, well intervention, flowback, and production. Besides well operations, the concept of SIMOPs also includes construction, commissioning, modifications, and maintenance. Examples of SIMOPs include the following [1]:

- In phased field development, production is started early while construction is still ongoing to add more production trains on the same well pad.
- The drilling crew works on a new well concurrently with production operations from existing wells.
- Modification to the production unit for debottlenecking, expansion, or regulatory compliance occurs without shutting down production operations in other trains on the same well pad.

Figure 1 Development Operations on a Typical Multi-well Pad in Shell Pinedale Shale Gas Field

Companies face significant challenges with unconventional shale gas projects to achieve their business goals while remaining focused on safety and minimizing impacts to the environment. Unlike continuous production from wells, many SIMOPs activities are temporary in nature and therefore bring an additional dimension of risk. Various simultaneously ongoing activities may place large numbers of personnel in close proximity to operating equipment in use for other activities. Activities such as construction or hydraulic fracturing may introduce a large number of ignition sources into the onsite area. It is therefore crucial to develop a comprehensive strategy to manage the SIMOPs.

Equipment layout and spacing is a fundamental aspect of risk management [2]. Proper siting and equipment layout is necessary to separate potential sources of fires, explosions, or toxic incidents from neighboring areas that could be involved in such events and protect people and assets from harm due to the events' potential consequences. In Canada, various provincial requirements or recommendations specify spacing for well site equipment, although the equipment is not specified for specific SIMOPs [3, 4]. In the United States, state regulations specify minimum spacing of permanent well site equipment from offsite facilities such as residential housing, public buildings, roads, and highways. However, to our best knowledge, there is no existing federal or state regulation for spacing and layout of temporary SIMOPs equipment on well sites in the United States.

Engineers in Shell's Technical Safety Engineering Team assisted Shell's unconventional gas projects and operations with identifying risks in the unconventional gas production business related to handling of flammable hydrocarbons and implementing effective measures to control the risks. In recent years, the Technical Safety Engineering Team has been helping various project and asset teams on fire hazard assessment for spacing and layout of SIMOPs equipment. The team recognized the need for a spacing guideline designed specifically to address the spacing problem associated with SIMOPs on shale gas facilities. In 2011, Technical Safety Engineering Team teamed up with Shell's unconventional gas business and Shell's Projects & Technology group to develop a new SIMOPs spacing chart that applies to all Shell unconventional shale gas well site locations in the United States. The purpose is to create a standard solution to a standard problem, to create more consistency in operations and elevate the bar for safety performance in Shell.

2. Project Overview

The project objective is to produce a set of recommended minimum spacing distances between SIMOPs equipment to prevent harm to people and damage to equipment from potential fire hazard due to hydrocarbon release. The chart is based on the minimum distances between flammable hydrocarbon sources and ignition sources. The chart will be applied in Shell unconventional gas development and production facilities in the United States.

Figure 2 depicts the three stages in this project:

1. Development
2. Review and Approval
3. Implementation

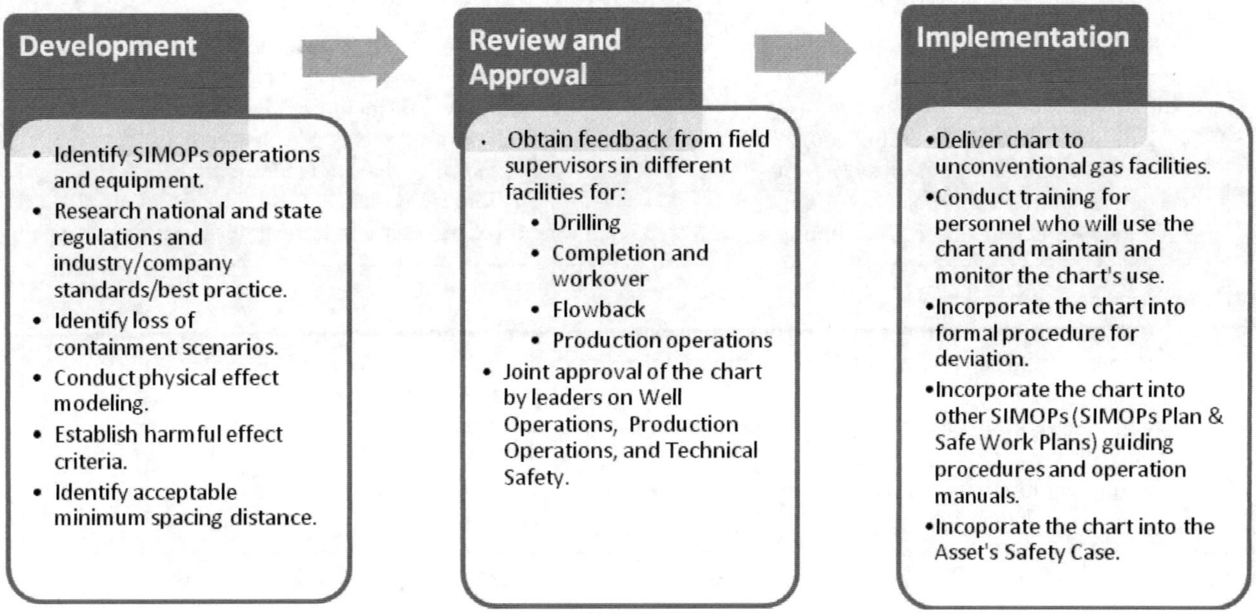

Figure 2 Project Road Map for Producing the New SIMOPs Spacing Chart

In the Development stage, the chart took its shape by following steps as listed in Figure 2. In the Review and Approval stage, the chart gained buy-in from its future users through the review and feedback process. In the Implementation stage, the chart was delivered and put into practice. Details of the stages are introduced in Section 3.

3. Development of SIMOPs Spacing Chart

3.1 Identify SIMOPs and related equipment

The first step in the chart development process is identifying the SIMOPs in Shell's shale gas well pad facilities. A cross functional project team that includes people from various backgrounds with knowledge and experience in SIMOPs is formed. The project team works together with operations superintendents, field supervisors, and wells engineers to identify a list of operations with associated equipment (Table 1) that will be covered by the SIMOPs spacing chart.

Table 1: SIMOP Equipment and Operations

Operations	Equipment
Drilling (excluding rig move-up and rig up operations) • Drill into reservoir, assessing composition of fluids in the rock for assessment of formation • Run various strings of casing • Cement to isolate different wellbore sections	Drilling well (including drilling rig) Drilling mud pits, shale shaker Diesel/oil-based mud storage tank Drilling rig boiler Mud de-gasser and atmospheric separator
Completions and Workover • Hydraulic fracturing: mix fresh or salty water with chemicals, add sand in blender, and pump fracturing fluid into well • Well intervention: perforating, wireline, snubbing, and coiled-tubing • Workover: stimulation of an existing production well to restore or enhance well production	Well under completion Completion high pressure equipment (including flow line, chokes, and trash catcher) Workover well (including service rig) Completion earthen pit
Flowback • Rig up well to temporary production equipment • Flow fluid in the well after a treatment to clean up the well and return the well to production	Flowback well Flowback separators and temporary piping Frac tank Flowback open top tank Temporary emergency flare tower, with engineered ignition source.
Production • Continuous flow of gas, with a certain amount of liquid (oil, condensate, or water) from well into surface equipment • Indirect heating through line heater • Three phase separation • Gas dehydration and compression • Chemical Injection • Liquid storage and loadout • Gathering and combustion of tank emissions • Inspection and maintenance work on production equipment	Producing well Flow line Production package (including indirect heaters, separators, dehydration, and metering) Instrument gas system Oil/condensate/produced water tanks Oil/condensate/water load-out tanker Chemical tanks Emergency relief systems Utilities (including heat medium reboilers)
Common • Operations that involve equipment which can be used in various phases of operations • Transportation on the well site • Construction: land preparation and vessel installation	Site offices and accommodation Fired equipment with/without flame arrestor Diesel/gasoline engines Hot oiler (propane fired) Flare tower Propane tanker for camp fuel delivery Well site LNG fuel supply and vaporizing equipment Designated smoking area Safe muster area

The SIMOPs equipment is further divided into two groups: potential release source of flammable hydrocarbons and ignition source.

Potential release source of flammable hydrocarbons

Equipment that contains flammable hydrocarbons is considered a potential release source. When loss of containment (LOC) occurs from such equipment, hydrocarbon is released into surrounding environment. Depending on properties of the released material and release conditions, there may be gas, liquid, or two-phase release. This document considers the following two forms of releases:

- Vapor release: gas dispersion and jet fire upon ignition.
- Liquid pool: formation of a liquid pool and pool fire upon ignition.

This document does not consider the scenario of pressurized liquid release and formation of liquid spray.

Ignition source

Equipment that may produce energy sufficient to initiate combustion in a flammable cloud is considered an ignition source. Ignition sources can be internal combustion engines, non-classified electrical equipment, or any fired equipment.

Based on this categorization, SIMOPs equipment is listed in the SIMOPs spacing chart, with equipment that is a potential release source of flammable hydrocarbons listed in the left column and equipment that is a potential ignition source listed on the top row. The distance in the block at the intersection of a row and column is the recommended minimum spacing for that pair of equipment (Figure 3).

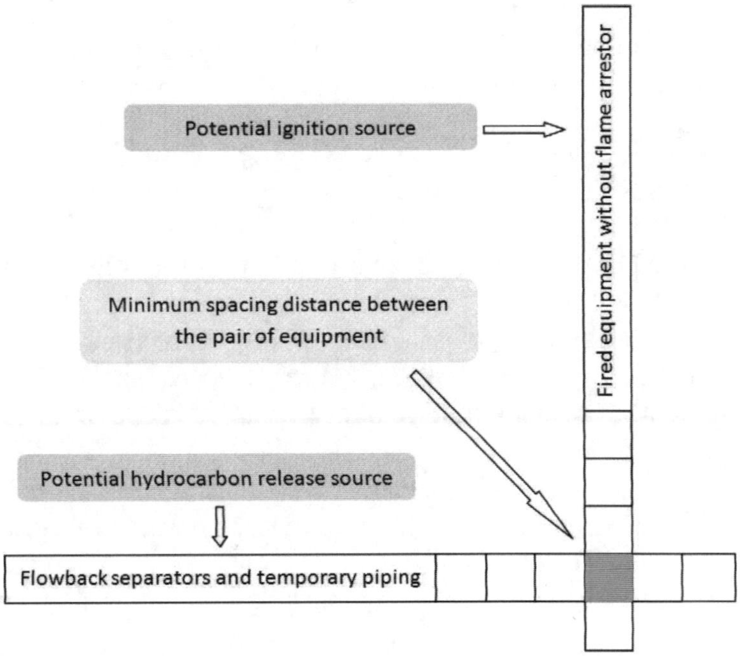

Figure 3 Minimum Spacing for Equipment Listed in SIMOPs Spacing Chart

3.2 Identify Loss of Containment scenarios

After the SIMOPs equipment is identified, the next step is identifying scenarios for how hydrocarbon leaks can occur and the consequences of such leaks. Identifying these release scenarios is a team workshop effort conducted using brainstorming to identify possible LOC scenarios for each type of SIMOPs equipment. Identified scenarios are classified into three categories: Major LOC, Most Likely LOC, and Credible Worst-case LOC.

Major LOC (M-LOC)

M-LOC is the worst-case possible release scenario that may happen. M-LOC involves the release of inventory in a large amount or an escape through a very large opening, which essentially establishes the metrics for the furthest extent for a flammable cloud or damaging fire heat flux. All M-LOCs are unintentional and undesired. Generally, an M-LOC is considered unlikely, and therefore is not usually applied as a design basis for SIMOPs spacing distances. Examples of M-LOCs include the following:

- Release from a full-bore ruptured pipe or a hole with a diameter of 150 mm or larger on a vessel.
- Release of gas from rupture or opening on a producing wellhead during loss of well control.
- Release of the entire liquid inventory from a condensate tank and ignition of vapor above the liquid pool.

Most Likely LOC (ML-LOC)

ML–LOCs are less severe release cases than M-LOCs. However, ML–LOCs are more likely to happen than M-LOCs. Examples of ML-LOCs include the following:

- Release of natural gas from a hole with a diameter up to 6 mm on high-pressure production equipment.
- Local minor release of natural gas from low-pressure or atmospheric systems.
- Local minor release of liquid hydrocarbons within bund area.

Credible Worst Case LOC (CWC-LOC)

CWC-LOCs are release cases that are less severe than worst-case M-LOC scenarios, but more severe than ML-LOCs. CWC–LOCs are usually based on the ML-LOCs, but with worse release conditions and more severe consequences, which are reviewed for conservative purposes.

- An example CWC-LOC scenario is natural gas release from a hole with a diameter up to 20 mm on flowback separator and associated piping. A maximum 20-mm-diameter hole size is adopted instead of a 6-mm hole for ML-LOC, because the temporary tubing on flowback equipment is considered more likely to have leaks than the permanent equipment for production.
- Another example is a case in which a loss of well control occurs after the fracturing operation, flow of gas from well was diverted to a temporary emergency flare tower, resulting in heat radiation from flaring.

Table 2 lists identified release scenarios using flowback equipment as an example.

Table 2 Examples of LOC Scenarios for SIMOPs Equipment

Equipment	Scenario type and description
Flowback equipment, separator, and production pack (including temporary flow lines and sales lines)	ML-LOC Scenario A gas leak occurs on temporary production equipment for flowback, such as the manifold, sand trap, fracturing stack, fracturing pipe, trash catcher, and temporary flow lines. A 10-mm diameter hole on a typical 3/8-in. pipe is assumed under 1440 psi pressure in the pipe.
	CWC-LOC Scenario A gas leak occurs on temporary production equipment for flowback, such as the manifold, sand trap, SCUD, fracturing stack, fracturing pipe, trash catcher, and temporary flow lines. Considering that leaks are more likely to occur on temporary pipes for the flowback, a larger hole size of 20-mm diameter is assumed for conservative purposes, under 1440 psi pressure in the pipe.
	M-LOC Scenario A full-bore rupture of the temporary tubing occurs due to pipe connection failure. The flow line carries mainly gas and flow rate is limited by the maximum output from the well based on a conservative estimate.

3.3 Select design-basis scenarios

The project team identified possible scenarios for each piece of equipment which contains hydrocarbon. Then among the scenarios for each piece of equipment, the project team selected one scenario as the design-basis case. Results of consequence modeling on the design-basis case will be used to determine the minimum spacing distance for the equipment.

Selection of the design-basis scenarios follows the general principle of Managing Risk in the Shell HSSE & SP Control Framework. According to the Shell HSSE & SP Control Framework, HSSE hazards in Shell businesses need to be identified and risks of the identified hazards need to be assessed using the Shell Risk Assessment Matrix (RAM; Figure 4). For hazardous scenarios classified as yellow and red cases on the RAM according to their severity and likelihood ranking, controls and recovery measures need to be identified and implemented to prevent the events from happening. The project team assessed each identified scenario with the RAM and identified corresponding controls and recovery measures by consulting field supervisors in Shell assets. Following the principle, the CWC-LOC scenarios were selected as the design cases for the majority of the SIMOPs equipment.

SEVERITY	CONSEQUENCES				INCREASING LIKELIHOOD				
	People	Assets	Environment	Reputation	A	B	C	D	E
					Never heard of in the Industry	Heard of in the Industry	Has happened in the Organisation or more than once per year in the Industry	Has happened at the Location or more than once per year in the Organisation	Has happened more than once per year at the Location
0	No injury or health effect	No damage	No effect	No impact					
1	Slight injury or health effect	Slight damage	Slight effect	Slight impact					
2	Minor injury or health effect	Minor damage	Minor effect	Minor impact					
3	Major injury or health effect	Moderate damage	Moderate effect	Moderate impact					
4	PTD or up to 3 fatalities	Major damage	Major effect	Major impact					
5	More than 3 fatalities	Massive damage	Massive effect	Massive impact					

Figure 4 Risk Assessment Matrix (RAM)

Although the likelihood of occurrence is lower for incidents with more severe consequences, detailed analyses on probabilities of the identified scenarios were not conducted during this project due to lack of release frequency data for onshore operations. Although risk-based methods are becoming more popular, the success of a risk-based analysis depends on a reliable and applicable database on probability of various release events. Accumulation of operational records and data in recent years means that leak frequency data are becoming more available and reliable for offshore operations. However, for onshore unconventional operations (especially the new shale gas development activities), available information is limited, making it difficult to conduct detailed probabilistic analyses for various events happening at onshore facilities.

3.4 Consequence modeling and selection of spacing distances

Physical effects from identified release scenarios in Section 3.2 were studied using a Shell proprietary software package for modeling the consequences, including flammable gas dispersion and fire heat radiation.

Based on the gas dispersion contour, a distance from the release source to gas concentration of lower flammability limit (LFL) is identified. The distance is used as the minimum spacing to prevent ignition of released gas. Based on the fire heat radiation contour, several distances are identified according to levels of the heat flux, each with different potentials to harm people or damage equipment. The heat flux levels are based on relevant Shell standards and industry guidelines. A distance from the release source to a heat flux level of 37.5 kW/m^2 is used as the minimum distance to avoid damage to equipment in a short-term period. A distance to 6.3 kW/m^2 is used to avoid injury to properly trained personnel during short-term exposure.

For each equipment, selection of minimum spacing distance depends on the objective of protection by spacing. Usually, spacing is intended to avoid ignition in the first place to avoid harm to people and damage to equipment. Therefore, the distance to LFL is used for most cases. For certain equipment, the distance to 37.5 kW/m^2 was used for conservative purpose (this is necessary for equipment whose failure may result in loss of critical functions and escalation of events during emergency situations, such as loss of well control). For certain equipment, protection of personnel from fire heat injury is necessary because operators need to work continuously in areas around the equipment. So the more conservative distance to 6.3 kW/m^2 heat flux level was used to allow personnel to escape from the fire. When selection of the distances is finished for all the SIMOPs equipment, the distances are entered into the spacing chart.

4. Implementation of SIMOPs Spacing Chart

4.1 Implementation across the organization

Success in any project cannot be achieved until the deliverable is effectively implemented. To successfully implement the new SIMOPs spacing chart, involvement of personnel in various levels of Shell organizations for unconventional shale gas projects at various locations is the most crucial part of the work. The people who were involved in chart development and who use the chart in their daily jobs are summarized below.

Wells Delivery Manager and Operations Manager

The Wells Delivery Manager is in charge of overall projects to deliver wells in a Shell unconventional shale gas field. The Wells Delivery Manager oversees operations, from drilling, completion, and hydraulic fracturing to later workover and well intervention jobs. The Operations Manager is in charge of producing activities in a field after wells have been completed. The Operations Manager oversees operations from initial flowback to bring the well online to daily production and maintenance, as well as later modifications to the existing facility, including abandonment. The Wells Delivery Manager and Operations Manager are the leaders who are accountable for safe operations in their assets. They need to be aware of the HSSE & SP risks associated with their business activities and how the risks are managed.
At the early stage of developing the new SIMOPs spacing chart, managers whose assets will use the chart were informed of the project, including the scope of the chart and form of deliverable. The asset managers understand the value of the chart as an effective risk management tool in the field and demonstrate their commitment to the project by assigning their field leaders and personnel to provide support and feedback.

Superintendents of drilling, completion, flowback, operation, and construction

Superintendents of drilling, completion, flowback, operation, and construction are responsible for specific operations in the field. For example, a drilling superintendent is in charge of all drilling activities at an asset. Superintendents are often the most experienced personnel in their areas of charged operations. Their knowledge and input are very valuable to the SIMOPs spacing chart, particularly when problems in the project cannot be solved by relying only on engineering calculations and industrial history and incident statistics on shale gas are limited.
During development of the chart, a team of superintendents from various assets was formed to discuss issues including equipment used in each operation, how things may go wrong, and potential hazardous scenarios. The continuous discussion helps establish the building blocks for the new SIMOPs spacing chart as described in the previous Section 3. Although it is difficult to achieve consensus on all the issues due to varied personal experience and understanding of relevant risks, the team aims to create a process in which everybody has a chance to provide

opinions for the work, and the team gains understanding of how and why a specific scenario was selected and a distance used.

Field personnel (supervisors, operators, HSE technicians, engineers, and foremen)

Field personnel are the main users of the SIMOPs spacing chart. For example, field supervisors need to consult the chart when reviewing or approving activities in advance of work. They need to make sure that the available onsite area is capable of accommodating the equipment in such a way that equipment spacing requirements will be followed, particularly when SIMOPs are occurring on the same pad.

Field engineers, operators and foremen need to consult the chart when there are changes in the work environment due to SIMOPs. They need to be aware of any change in equipment layout, identify any spacing issues that cause concerns for the safety of workers and facilities, use the chart to assess the conditions, and raise issues to supervisors. Field HSE technicians need to make sure that a copy of the chart is available to onsite personnel and is followed constantly.

4.2 Integration with overall SIMOPs risk control system

The SIMOPs spacing chart is a tool designed to prevent ignition of hydrocarbons from potential fixed and temporary ignition sources. When using the chart, it is important to recognize that there are many other hazards besides flammable hydrocarbon and fire, such as physical exposure from high pressure in temporary piping, dropped object, and collision. There are also many other Shell risk mitigation measures and tools to manage risks in SIMOPs. These measures constitute a system within which the new SIMOPs spacing chart should be used. These measures are briefly introduced below.

Manual of Permitted Operations (MOPO)

MOPO is a visual tool developed by Shell Technical Safety Engineering Team in coordination with relevant Shell organizations [5]. MOPO brings together a collection of field activities and situations into a matrix sheet and provides information about whether certain combinations of SIMOPs activities can add risks to safe operations.

MOPO originated from findings by work planners and supervisors when they were reviewing work permits offshore. It is a valuable tool in providing a means to assist with consistently applying risk management controls to the various SIMOPs activities occurring on a daily basis in offshore platforms. Shell started to implement MOPO to its offshore installations in the Gulf of Mexico in 2007. Based on the successes in the Gulf of Mexico, MOPO for onshore business was developed and has been applied in onshore field production facilities, gas plants, and in-situ oil production in Canada and the United States since 2009.

MOPO uses a traffic light chart format to indicate when SIMOPs activities should not proceed (red), proceed with caution (yellow), or proceed without special restriction (green). For the activity combinations marked with yellow, SIMOPs should be carried out when certain restrictions are complied with. Among the 12 listed restrictions, 2 refer to the spacing chart:

- Restriction #1: as long as activity is not close to other activity, judgment of Person in Charge (PIC)/Supervisor/Spacing Chart.
- Restriction #4: wells within vicinity to be temporarily shut in, judgment of PIC/Spacing Chart.

Field supervisors must use the new SIMOPs spacing chart together with MOPO to assess and approve SIMOPs subject to the restrictions on equipment spacing.

Wells Worksite Instruction Manual (WWIM)

WWIM contains a summary of key company standards that are deemed the most critical for safe and efficient delivery of well operations. WWIM identifies 10 areas for which the mandatory requirements are laid out, including Permit to Work, Lifting and Hoisting, Dropped Object Prevention Scheme, temporary pipe work, and emergency preparedness. Among the requirements are that personnel exclusion zones shall be established, properly barricaded, and access-controlled to prevent physical pressure hazards and dropped objects. Because certain SIMOPs equipment must be continuously manned or constantly checked on, it is important to consider the necessary personnel exclusion zone or extra spacing distance when planning spacing of the equipment, besides using the SIMOPs spacing chart for prevention of fire hazards.

Sour Gas Protection

Unconventional shale gas reservoirs may contain high concentration of hydrogen sulfide (H_2S or sour gas). As an acute toxic substance, H_2S has caused fatalities in the oil and gas industry, including the Chuandongbei gas well blowout in China in 2003, which killed 243 people and injured more than 9000 people. In Shell, managing risks due to exposure to H_2S is listed as a Process Safety Basic Requirement (PSBR) in the HSSE & SP Control Framework, which requires assessment of safety risks in H_2S operations, particularly for concurrent operations or SIMOPs.

Shell International Exploration and Production developed a special concept called Time to Protect as a layout methodology for SIMOPs in high-pressure and high-H_2S-concentration facilities [1]. According to the Time-to-Protect methodology, when SIMOPs activities are ongoing in areas nearby a facility where H_2S is present, a safe separation distance must be established to allow safe evacuation of personnel upon early detection of an H_2S release. Therefore, larger equipment spacing might be needed based on this methodology.

Because the SIMOPs spacing chart is designed for "standard" or sweet gas facilities, facilities where H_2S is present must follow relevant Shell procedures to assess layout and equipment spacing to take H_2S presence into account, such as using dispersion models to determine the dimensions of the 300-ppm H_2S plume from credible release events and locating the SIMOPs work area outside of the plume's footprint.

5. Summary

A spacing chart for SIMOPs was prepared for unconventional shale gas development and production fields in Shell in the United States. Development of the chart provides an opportunity to check through onsite risks in a systematic way by identifying various scenarios that can result in flammable hydrocarbon leak and fire. Valuable experience and knowledge on field operations and risk management is gained during the process. The chart's implementation process enhances awareness of onsite risks through communication and training across Shell organizations and locations. The SIMOPs spacing chart is an important part of the overall system to manage risks from SIMOPs in Shell's unconventional onshore assets in the United States.

6. Acknowledgement

Certain key people enabled success in this project. Support from the Operations and Wells Delivery Managers in Shell Exploration and Production Company is appreciated, especially because any changes and improvements in safety can only be realized with demonstrated leadership.

We are also grateful for the support, time, and valuable input of the following members of Shell Technical Safety Engineering Team: Douglas Detman, Sam Sheldon, Matt Childs, Melissa Velasquez, John Pruitt, Jennifer Morgan, and Jeff McPhate.
Finally, we thank all the field personnel, superintendents, and operators for their support and guidance during this project.

7. References

[1] Blyth, A., "Time to Protect: Layout Methodology for Safe SimOps in High-Pressure, High-% H_2S Facilities", SPE Americas E&P Environment & Safety Conference, San Antonio, Texas, USA, March 23-25, 2009.

[2] CCPS. *Guidelines for Facility Siting and Layout.* Center for Chemical Process Safety. AIChE, 2003,

[3] British Columbia Oil & Gas Commission, 2013, Well Completion, Maintenance and Abandonment Guideline, April 2013.

[4] ENFORM The Safety Association for Canada's Upstream Oil and Gas Industry, 2008, Industry Recommended Practice (IRP) Volume 20, Wellsite Design Spacing Recommendations.

[5] Detman, D., Groot G., "Shell's Experience Implementing a Manual of Permitted Operations", 14th Annual Symposium, Mary Kay O'Connor Process Safety Center, Texas A&M University, College Station, Texas, October 25-27, 2011.

Technology Validation Process for Offshore Fields: A Winding Road

Ankur Jariwala

Senior Product Manager – Gas Technology
Cameron Process Systems, Houston

Gary W. Sams

Managing Director R&D
Cameron Process Systems, Houston

The technology qualification process is usually a winding road to commercialization. In this paper we will discuss Cameron's experience in working with companies to qualify gas technology for their offshore and onshore commercial operations. In last 30 years, we have commercialized over 18+ gas technologies and have gone through significant internal and external technology qualification steps starting from concept validation, bench scale testing, commercial approach validation, pilot testing, engineering, design and installation. The process can vary significantly from onshore to offshore. We have a wide range of technology validation experiences with technologies targeted for offshore, mainly Cynara CO_2 Membranes, MEG Reclamation and Regeneration technology, electrostatic oil dehydrators and high efficiency phase separation internals. Although the external technology validation process may vary, there are several common aspects to this process. Common aspects are operational history in a similar environment, mean time between failures, cause and effect chart, mechanical and structural integrity, process variations, utility consumption, ongoing research and development and aftermarket support.

The technology validation process also requires integrating different Original Equipment Manufacturer (OEM) and associated technology validations to align with parent technology. This is also important to identify, define and incorporate into an overall technology validation strategy for the technology success. In this presentation, we will also cover several challenges and hurdles we faced with OEMs and their associated technology and how we overcame those risks via a technology validation process.

Technology evolution is a process where technology runs through a maturity phase via prototype, bench scale, pilot unit and a semi or a full commercial unit operation. Each of these steps provides valuable data to main commercial unit and there by minimizes the overall risk from laboratory model to commercial scale unit. Especially Oil and Gas patch, where technology has to be robust due to potential safety and failure risks which can drive significant revenue loss. Key measures are put in place to validate the technology before applying to the main field.

For offshore oil and gas production, the technology validation process is more rigorous compared to onshore applications. Cameron has developed several oil and gas processing technologies for the offshore which has gone through technology validation process. The four main technology validation process experiences listed here shows different sets of processes than any standardized procedure. However, they all have one thing in common, long term offshore testing to analyze technology uptime and its operational success under a tough upstream environment.

Technology ideation starts with a problem statement or a pain point. The problem statement creates multiple options or ideas for the solution. Ideas can be easy to implement when it's a proven technology in the respective field. The available technology can be in the parallel field and may be required to go through a qualification process for the required field. The technology qualification process is identified as technical as well as commercial. The technical qualification can run through different stage gate process which includes an assessment of the Technology Readiness Level (TRL). The TRL levels usually run from 1-9 and depending upon how the technology is proven and tested, it's ranked accordingly. Most of the times, companies/clients would like to investigate technology once it has reached to level 3-4. Again, it may vary depending upon how difficult the problem it's going to solve and what is the biggest impact the technology can have, even if it is at TRL level 1 or 2.

The commercial qualification is usually evaluated once the technology has reached TRL level 4-5. The commercial validation process really focuses on Return on Investment (ROI), technology risk, potential market penetration versus cost and time to market. Each of these factors is weighted for the given technology to get a combined commercial qualification score.

The process of technology qualification can be application specific, which can be completely developed and proven by the technology owner. However, the technology owner may choose to take the technology to where the industry or end users have enough data and comfort level to accept technology for the required problem statement. In some cases, end users have more interest and may take the full responsibility of the technology validation or testing of the technology for the required application.

Technology qualification can also be done through a third party. For example, a technology owner can choose to use innovative technology from the third party as a part of their main technology. This process is also more like Original Equipment Manufacturer (OEM) where the component of the main technology is unique and may require the same qualification process as main technology. In this case, technology owner can work with third party or OEM provider to run in a same TRL or stage gate process which is parallel to main technology qualification.

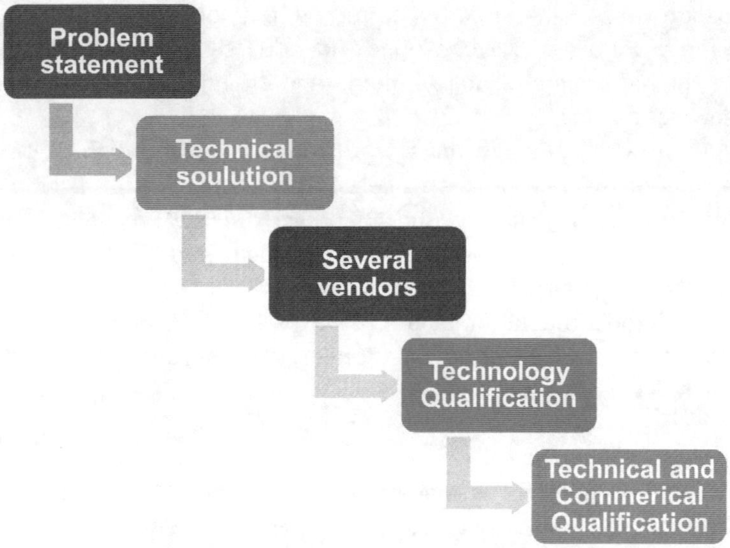

Figure 1.0 – Technology Qualification Process and Various steps

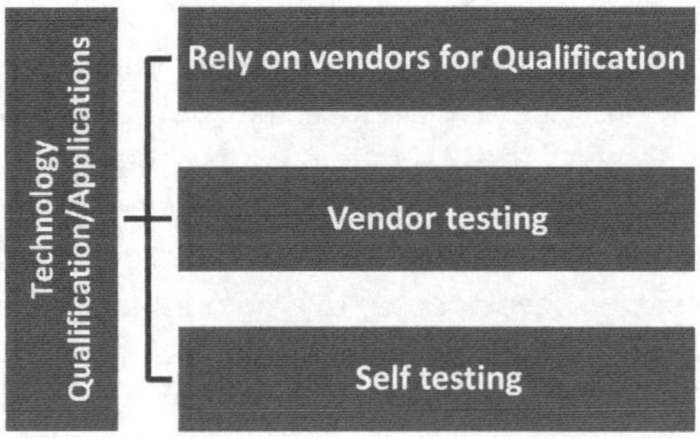

Figure 2.0 – Technology application qualification options

The four cases below expand on the Technology Qualification methods used by Cameron to qualify technology for a range of oil and gas processing applications.

1) Cynara Membrane System – Technology validation process

Problem:

Membrane for separating CO_2 from natural gas is a technology which is very well proven for onshore gas processing. For the offshore, the system footprint has to be small and should be able to address the change in inlet feed gas composition in the dynamic manner. The onshore CO_2 removal using membrane systems have limited control system in past and cannot go full automation which is required by the remote offshore gas fields. For the offshore fields there were several challenges to address like mean time between the failures, higher uptime reliability, membrane change out, sparing philosophy and turn down capability.

Most of the offshore units are designed with the highest design capacity as per life of the field to plan overall space and weight on the platforms or Floating Production, Storage and Offloading (FPSO) vessels. During the first year, the production may not see the full capacity and therefore need a design to address that turndown.

Cameron's recent projects on FPSO require compact footprint and higher automation to minimize operator attention. These challenges require different problem solving approach.

Technology Solutions:

Cameron has proven membrane technology, Cynara membranes, with 28 years for operational experience onshore. Cameron's extensive experience in operating some challenging Enhance Oil Recovery (EOR) gas fields with high CO_2 at medium to high pressure on an ongoing basis provides significant operational data for hollow fiber membranes. Cameron's Cynara membranes use Cellulose Tri Acetate (CTA) polymer based hollow fiber membrane which are well proven. For small footprint requirements offshore, Cameron never changed their core fiber and stuck to CTA material based membrane while improving thee overall packaging method to reduce the overall size and weight of the skid.

Cameron started with 5" diameter membrane module using the same CTA membrane fibers and developed a 16" membrane diameter element which can replace 18 of the 5" membrane elements. As the offshore fields are getting larger with higher CO_2, the approach is to reduce this footprint even further by developing a 30" diameter membrane module. Each 30" membrane module can replace 4 of the 16" diameter membrane elements, significantly reducing space, weight and system cost.

Figure 3.0 – Different Cynara Membrane Module Sizes.

The 30" diameter membrane provides significant footprint reduction with higher uptime.

Technology Validation Process:

The 30" diameter membrane element is about 72" tall and has more capacity to withstand a higher pressure with higher hydrocarbon retention rate in the product gas.

These membrane modules were not tested offshore and our client was a bit concerned over the performance of these modules, because they were going to put them in a main service where they play critical role in their revenue chain. The first approach is to test this in a real process pilot unit at SACROC; Cameron's operated membrane based CO_2 removal facility in West Texas. Cameron developed the data set range for the client who was interested in testing this offshore at their facility. The process data for the 30" membrane element showed predictable performance under desired conditions for 3 years of testing at SACROC. After reviewing these data, the client agreed to work with Cameron in testing this 30" membrane module at their offshore facility. They started the offshore test and ran for another 3 years to understand its performance under variable feed gas conditions.

During this period, Cameron had developed and improved their control system to minimize any downtime and developed better sparing philosophy by working with the client.

After 3 years of testing offshore, the client qualified this technology for their further projects for onshore, as well as offshore.

2) MEG Reclamation and Regeneration Technology

Problem:

The technology was very well qualified for the given field, however the client's application was unique and required testing and qualification for the major capital investment decision.

One of the Cameron clients approached them regarding their existing problem at their gas terminal in UK. The facility was developed long time ago and it is currently processing gas more than their original design conditions. This was a major revamp which the customer was looking to do in order for them to operate this facility efficiently.

In their facility, monoethylene glycol is injected into the gas production pipelines to prevent formation of gas hydrates in the raw gas at pipeline conditions (high pressure and low temperature). The facility includes MEG Regenerators to remove the produced water from the incoming Wet MEG feed (50 wt% MEG) to yield Dry MEG by removing water.

Technology Solution:

A Desalination Package was installed during the earlier phase of production to remove salt from the Dry or Wet MEG. Both monovalent salts (predominantly sodium chloride) and divalent salts (calcium chloride) were present in the feed. The divalent salts are removed as insoluble carbonates in a Pre-Treatment Package and the monovalent salts are precipitated from the Dry/Wet MEG in the Desalination Package using flash vaporisation.

The Pre-Treatment and Desalination Packages were to be replaced as part of the Rejuvenation Project. Client's production projection for that gas fields indicated increased water and salt production rates.

Technology Validation:

In 2010, the Client engaged Cameron to carry out a FEED study for the Desalination Package with a new feed design basis.

As part of the FEED study, pilot plant trials were carried out at Cameron's PureMEG™ pilot plant at the University of Manchester in 2012 using Dry MEG supplied by Client from the existing Desalination Package from their field. The pilot plant trials were carried out to demonstrate Cameron's proprietary salt clean-up technology to Client's process engineering, projects and operations personnel. In Cameron's PureMEG™ process, MEG is washed from the salt crystals as they settle under gravity through a column of saturated sodium chloride brine. The result is reduced MEG losses with the salt waste stream. The PureMEG™ Pilot Plant comprises a 600mm ID x 1.6m high flash separator above a 6 m high, 4" diameter column fabricated from borosilicate glass which allows observation of the salt crystals falling through the brine phase.

Figure 4.0 – Cameron's PureMEG Pilot Unit at University of Manchester, UK

The pilot plant trials were also used to demonstrate (1) first fill of the system with MEG, (2) control of the brine column during salt processing and (3) the robustness of Cameron's flash separator design to the presence of Flow Assurance Chemicals (particularly corrosion inhibitors) which cause foaming problems on the existing unit.

The pilot plant trials were carried out over a period of 5 days and were witnessed by the client.

In addition to the Flash Separator trials at the University of Manchester, trials were carried out with filter press and centrifuge vendors to support the Client's Rejuvenation Project FEED study.

After several testing and trial runs, the Client was comfortable with the technology and awarded the contract for the design of the Desalination Package of the Rejuvenation Project to Cameron Construction / fabrication and commissioning is due to begin in 2014.

Again, the technology was very well proven; however, it still does not meet the required criteria from the Client unless they test this technology and get the detailed performance data. The approach is to really minimize the risk with the technology by investing more time upfront by qualifying any technology at their cost.

3) Dual Frequency Electrostatic Dehydrators

Problem:

When applying electrostatic processes to a crude oil for the purpose of oil dehydration, it became apparent that the electrical conductivity was compromising the process performance. The problem statement posed the question of determining a method that would permit a high voltage electrostatic field to be sustained even in the presence of a highly conductive crude oil. The existing technology was being compromised by the excessive current demand on the power supply and subsequent loss in voltage. Clients were experiencing poor dehydration due to the reduce voltage and loss of dehydration efficiency.

Technology Solution:

The technology solution was found by investigating the use of medium frequencies that could successfully sustain the voltage field. The application of a medium frequency power unit proved to be a valid method to achieve the desired dehydration from highly conductive oils. During the development of this solution it was discovered the conductivity resulted primarily from the blending of two oils that chemically reacted causing the resulting conductivity to spike.

The function and features of the medium frequency power supply were clearly defined to facilitate the work scope.

Technology Validation:

Using the function and features documents, the power unit was designed and fabricated for pilot testing on crude oil samples supplied by potential end users. Numerous oils were tested using the pilot power unit to demonstrate the benefit of this new technology and to validate and confirm the functionality. Following several months of pilot trails and a few design changes to the power unit a commercial power unit was design and built. The commercial unit was tested on a pilot plant before going for a field trial.

Figure 5.0 – Electrostatic treater in the field

Several oils were tested and a few features were altered in the power unit before looking for a field site.

The first commercial qualification by a customer was conducted at a large production facility operating five oil dehydrators in parallel. One dehydrator was converted to the new power unit and the field trials were started. Obviously, in a commercial facility the ability to alter performance parameters is typically limited to reduce the risk of loss production. For these trials, the oil capacity was the primary test variable. The rate was increased in several steps and the dehydration performance of the equipment was recorded. Once the rate tests were completed, the unit was permitted to operate without intervention to permit the equipment to handle the daily upsets experienced by the process. After 30 days of operation, the power unit was removed and shipped to the manufacturer for examination. These field trials resulted in a few final alterations to the design.

These field trials provided the needed testing to satisfy a TRL of 5 for this technology making it readily acceptable by many operators.

4) High Efficiency Separator Internals:

Problem:

Oil, gas, produced water and several other phases are separated in a separator via gravity separation. These phase separators are sometimes not efficient enough to meet the separation requirements due to inadequate distribution of phases inside the separator or due to a change in the inlet flow conditions.

Technology Solution:

High efficiency Separation internals are used in phase separation/separators where oil/gas/produced water and solids are separated to the highest efficiency and thereby protecting downstream equipment from any type of two phase handling. These internals are developed to either retrofit an existing separator to increase the overall throughput while maintaining efficiency, or they are installed in a new separator to meet highest level of separation performance.

The technology was initially developed as per the required standards from offshore projects. Cameron developed these cyclonic separator internal designs and developed some prototypes in the beginning based on the concept and computational fluid dynamics study.

For the offshore operational environment, these CFD analyses are not sufficient enough to qualify the technology. Our client, who is a highly technology oriented company, requires certain testing protocol before they can qualify these internals for their commercial separators offshore.

Technology Validation:

The protocol called for an extensive CFD analysis with various flow conditions and required certain changes in the physical structure of these internals. After rigorous analysis and satisfactory preliminary results, these internals were tested on a bench scale unit in-house to understand the basic performance in an air/water separation regime. Some limitations on the bench scale were the test pressure and actual fluid conditions which are different from oil and gas field. After successfully testing and understanding the operational envelope for these internals, the internals were tested at very high pressure and actual field simulated conditions at the client's laboratory. These tests provided complete real time performance and reduced the overall performance risk on new technology. This also provides technology limitations and operational efficiency envelope under different process flow conditions.

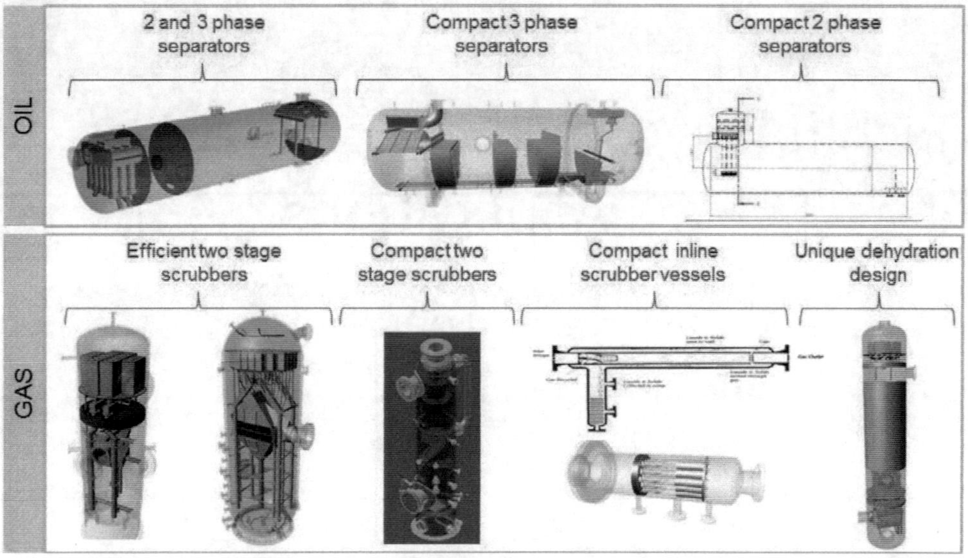

Figure 6.0 – Cameron's High Efficiency Separator Internals Suite

Cameron has successfully developed 5 different types of internals which are targeted for different gas loading factors. All of these internals were tested and approved by the Client. One of the internal designs was patented by the client and Cameron was exclusively authorized for licensing. The Client has full access to the internals for their existing or future projects.

By working with a client on technology development and qualification process, the technology accessibility becomes much easier for their key technical team and it also reduces the overall technology risk for the client.

Conclusion:

As offshore oil and gas production fields are getting more complex, they require more sophisticated technologies to produce cost effectively. The technology validation process is a key aspect to address safety and reliability. The technology validation process may vary significantly depending on the type of the application. The overall process requires close cooperation between technology user and vendor. It also requires strong research and development capability to validate technology from laboratory scale to fully commercial technology.

Cameron has developed extensive and successful experience in validating new technology for large complex offshore fields with significant research and development capability in house.

Technology Qualification – Two Subsea Applications

Martha Viteri – DNV-GL

Austen Tyler – DNV-GL

14 AIChE Spring Meeting, New Orleans, LA

DNV·GL

Technology Qualification

- Qualification is the process of providing the **evidence** that technology will **function within specified limits** with an acceptable level of **confidence**.

14 AIChE Spring Meeting, New Orleans, LA

DNV·GL

Why Consider Technology Qualification?

- Lack of evidence in the technology
- Internal or customer requirements
- Technology selection
- Expert evaluation
- Market competitive advantage

14 AIChE Spring Meeting, New Orleans, LA

DNV·GL

Technology Qualification Process

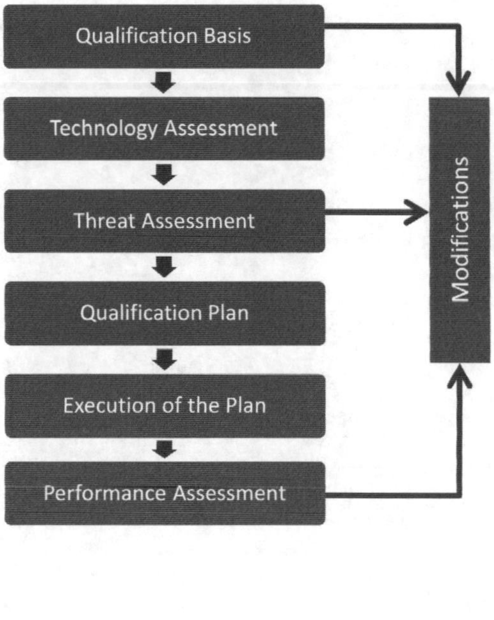

14 AIChE Spring Meeting, New Orleans, LA

DNV·GL

Qualification Basis

- Common set of criteria to assess qualification activities
- Describe the following:
  - Intended use
  - Environment
  - Required functions
  - Acceptance criteria
  - Performance expectations

14 AIChE Spring Meeting, New Orleans, LA

DNV·GL

Technology Qualification Process

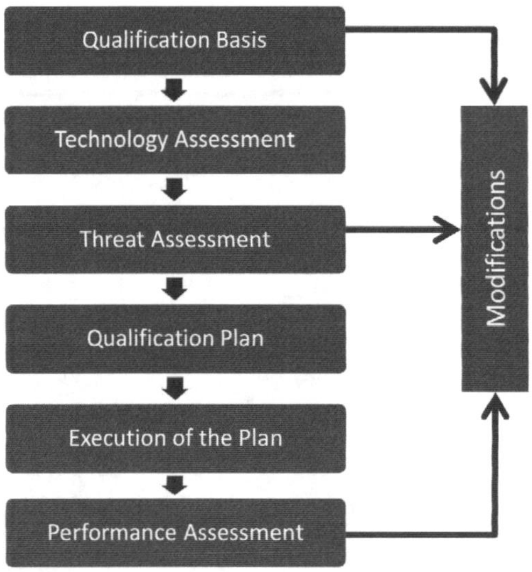

14 AIChE Spring Meeting, New Orleans, LA

DNV·GL

Technology Assessment

- Identify which elements require technology qualification
- Identify key challenges and uncertainties.

14AIChE Spring Meeting, New Orleans, LA

DNV·GL

Technology Assessment

- Consists of the following steps:
  - Breakdown the technology
  - Assess technology novelty
  - Identify key challenges and uncertainties

14AIChE Spring Meeting, New Orleans, LA

DNV·GL

Technology Assessment

Application Area	Novelty of the Technology		
	Proven	Limited Field History	New or Unproven
Known	1	2	3
Limited Knowledge	2	3	4
New	3	4	4

- The classification categories:
  1. No new technical uncertainties
  2. New technical uncertainties
  3. New technical challenges
  4. Demanding new technical challenges

14 AIChE Spring Meeting, New Orleans, LA

DNV·GL

Technology Assessment

- Statement of Feasibility:
  - Document issued by DNV-GL
  - Affirms technology feasibility
  - Based on DNV-RP-A203

DNV

DET NORSKE VERITAS

STATEMENT OF FEASIBILITY

<Technology Name>

Technology owner: <Company Name>

Name of technology: <Technology Name>

Description:

Application:

Involvement:

Limitations:

Reference documents:

<Place and date>

for Det Norske Veritas AS

Business Development Leader

Project Manager

14 AIChE Spring Meeting, New Orleans, LA

DNV·GL

Technology Qualification Process

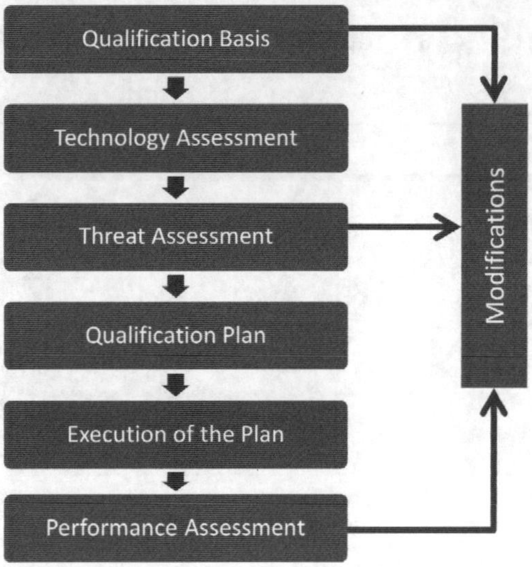

14 AIChE Spring Meeting, New Orleans, LA

DNV·GL

Threat Assessment

- Identify failure modes / mechanisms
- Assess the associated risk
- Most common methodology chosen is FMECA

Failure Mode	Failure Mechanism	Failure Effect	Failure Detection	Safeguards	Risk Ranking
Mode 1	Mechanism 1	Effect 1			High
	Mechanism 2	Effect 1			Low
Mode 2	…				

14 AIChE Spring Meeting, New Orleans, LA

DNV·GL

Threat Assessment

- Risk ranking performed using a risk matrix

		Increasing Likelihood				
		1	2	3	4	5
Increasing Severity	1	L	L	L	M	M
	2	L	L	M	M	M
	3	L	M	M	M	H
	4	M	M	M	H	H
	5	M	M	H	H	H

14 AIChE Spring Meeting, New Orleans, LA

DNV·GL

Technology Qualification Process

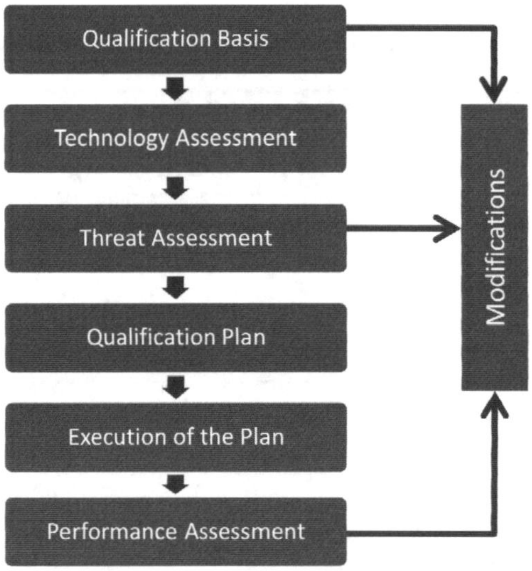

14 AIChE Spring Meeting, New Orleans, LA

DNV·GL

Qualification Plan

- Evidence needed to manage the critical failure modes
- Describe the following:
  - Evidence linked to failure mode
  - Qualification activity linked to Qualification Basis
  - Success criteria

14 AIChE
Spring Meeting, New Orleans, LA

DNV·GL

Qualification Plan

- Potential qualification methods include:
  - Engineer judgment
  - Analytical methods
  - Numerical methods
  - Testing
  - Procedure development

14 AIChE
Spring Meeting, New Orleans, LA

DNV·GL

Qualification Plan

- Endorsement of Qualification Plan:
  - Document issued by DNV-GL
  - Affirms ability to qualify technology through the planned activities

DET NORSKE VERITAS

ENDORSEMENT OF QUALIFICATION PLAN

<Technology Name>

14 AIChE Spring Meeting, New Orleans, LA

DNV·GL

Technology Qualification Process

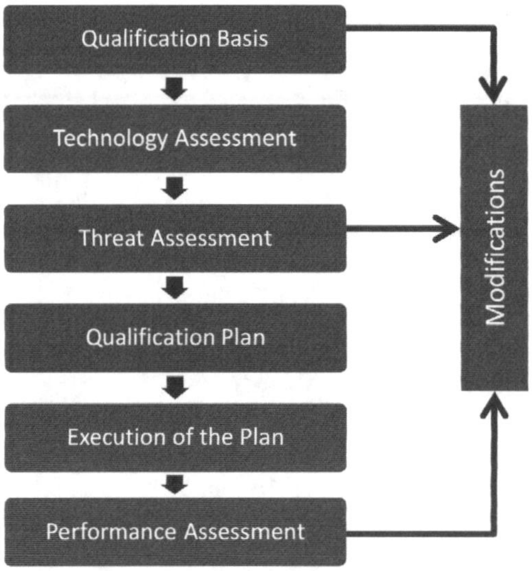

14 AIChE Spring Meeting, New Orleans, LA

DNV·GL

Qualification Plan Execution

- Involves the following steps:
  - Execution of qualification plan
  - Collecting and documenting data generated
  - Ensuring traceability of the captured information
  - Determining the performance margin

14AIChE
Spring Meeting, New Orleans, LA

DNV·GL

Technology Qualification Process

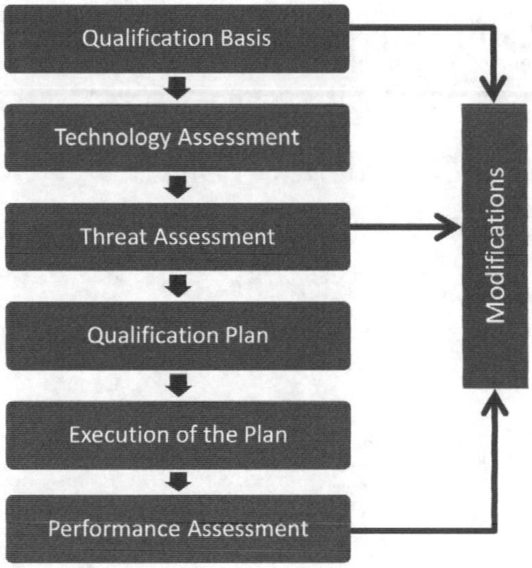

14AIChE
Spring Meeting, New Orleans, LA

DNV·GL

Performance Assessment

- Confirmation that requirement have been met
- Risk and uncertainty reduced to acceptable levels

Parameter	Acceptance Criteria		Failure ID	Qualification Activity	Margin	
	Min	Max			Min	Max
Number of Cycles	100		2.5	Five samples will be cycled until failure	50	
…						

14 AIChE Spring Meeting, New Orleans, LA

DNV·GL

Performance Assessment

- Technology Certificate:
  - Document issued by DNV-GL
  - Based on Qualification Basis and relevant standards
  - Affirms qualification according to DNV-RP-A203

DET NORSKE VERITAS
TECHNOLOGY CERTIFICATE
<Technology Name>
This Certificate is valid until

14 AIChE Spring Meeting, New Orleans, LA

DNV·GL

Qualification Progress by Development Phase

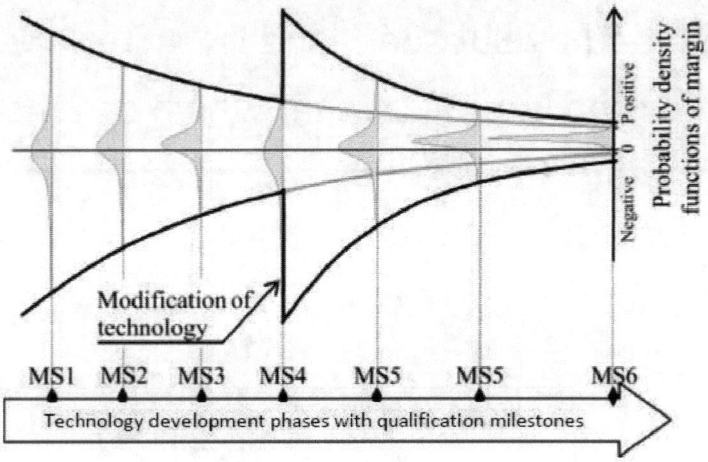

14AIChE Spring Meeting, New Orleans, LA

DNV·GL

Enhanced Thermal Performance Reeled Pipe-in-Pipe System

- Challenge:
  - Maintain the appropriate product temperature
  - Avoid the formation of hydrates or waxing

14AIChE Spring Meeting, New Orleans, LA

DNV·GL

Enhanced Thermal Performance Reeled Pipe-in-Pipe System

- Developed by Subsea 7 and ITP InTerPipe
- Reeled pipe in pipe system which can include electrical trace heating
- Stepped approach to qualification

14 AIChE Spring Meeting, New Orleans, LA

DNV·GL

Reelable BuBi® Lined Pipe System

- Challenge:
  - Transporting corrosive media
  - Develop lower cost pipeline systems

14 AIChE Spring Meeting, New Orleans, LA

DNV·GL

Reelable BuBi® Lined Pipe System

- Developed by Subsea 7 and Butting
- Without the occurrence of any wrinkling or other degradation of the corrosion-resistant liner, via reel-lay method.

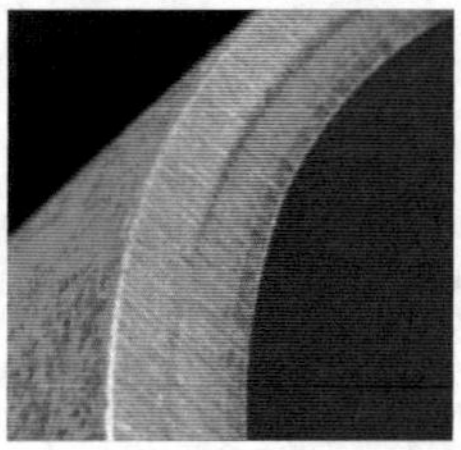

14AIChE Spring Meeting, New Orleans, LA

DNV·GL

Outcome of the TQ Projects?

- Evidence / confidence in the technology
- Understand risk profile and failures
- Compliance with requirements
- Independent evaluation of the technology
- Market competitive advantage

14AIChE Spring Meeting, New Orleans, LA

DNV·GL

Closing

- The authors would like to thank Subsea 7, ITP and Butting for their permission to publish
- austen.tyler@dnvgl.com
- Any Questions?

14 AIChE Spring Meeting, New Orleans, LA

DNV·GL

Well, Well!
Are We Drilling in the Right Direction?

Michael P. Broadribb
Baker Engineering & Risk Consultants, Inc.
3330 Oakwell Court, Suite 100, San Antonio, Texas 78218-3024
mbroadribb@bakerrisk.com

Prepared for Presentation at
American Institute of Chemical Engineers
2014 Spring Meeting
10th Global Congress on Process Safety
New Orleans, LA
March 30 – April 2, 2014

UNPUBLISHED

Well, Well!
Are We Drilling in the Right Direction?

Michael P. Broadribb
Baker Engineering & Risk Consultants, Inc.
3330 Oakwell Court, Suite 100, San Antonio, Texas 78218-3024
mbroadribb@bakerrisk.com

Keywords: offshore drilling, process safety, human factors

Abstract

From the Piper Alpha to the 2010 Deepwater Horizon/Macondo incident and many others in between, offshore oil and gas operations clearly have the potential for catastrophic consequences. Although most of these incidents involved a number of technical causes, many of the root causes were related to human factors involving leadership, accountability, and safety culture issues.

Individual performance can be negatively affected by factors such as motivation, fatigue, inadequate training, excessive workload, poor equipment design, and the worksite environment. Most industrial incidents involve human factors, and understanding these performance-shaping factors is essential for learning lessons and preventing recurrence. A strong culture takes advantages of inherent strengths and helps overcome weaknesses of each individual within the company to create optimal company performance.

The author recently interviewed onshore teams responsible for well design and execution of drilling and completions (D&C), and the crews of mobile offshore drilling units (MODUs) and drilling facilities on production platforms. Observations of these D&C operations in different geographical regions and different regulatory regimes - including the Gulf of Mexico, North Sea, West Africa, and Asia Pacific - highlighted a number of common issues with potential to adversely impact safety.

1. Introduction

Today, many offshore drilling and completions (D&C) operations clearly have the potential for a major incident, such as loss of well control, with similar exposures as the 2010 Deepwater Horizon incident [1, 2]. Although there were a number of technical causes of this incident, many of the root causes appear to be related to human factors involving safety culture and leadership.

Over the past 5 years the author has had the opportunity to observe and interview D&C teams from several oil companies, the crews of multiple mobile offshore drilling units (MODUs), production platform-based drilling facilities, and onshore drill rigs operating around the world. This has given the author insights into strengths, weaknesses and common themes involving human factors, safety culture, and the effectiveness of management systems for process safety[1] and asset integrity. Variations in culture and practices between different regions and different regulatory regimes also influence process safety performance.

Comparisons can be drawn to the expert testimony that the author and others presented at the Piper Alpha inquiry, and to the recommendations made by the inquiry chairman, Lord Cullen, and his advisors 25 years ago [3]. The testimony and recommendations were aimed at preventing future major incidents in the North Sea, but inevitably had an impact on oil and gas exploration and production operations around the world. Twenty-five years later, some of those lessons around leadership, risk assessment, permit to work, and emergency response are equally applicable today. However, it sometimes seems that the lessons have been forgotten or ignored and today's generation is destined to make the same mistakes and learn the hard way.

2. The Legacy of Piper Alpha

On July 6, 1988, a major disaster in the North Sea took the lives of 165 persons on board the Piper Alpha oil production platform and 2 of the rescue crew from the Sandhaven, a standby vessel for a nearby MODU. To this day, the incident represents the worst offshore oil industry disaster ever, and has become an industry changing watershed event.

The subsequent public inquiry report (known as the Cullen Report) [3] made a number of recommendations related to inherently safer design (ISD) and, as a result, future production platform designs invariably have a rectangular plot plan/layout rather than the Piper Alpha's essentially square layout. The recommendations from the Cullen Report included the mandatory use of a systematic fire and explosion analysis, analysis of smoke and gas ingress to living quarters, and the requirement for a temporary (safe) refuge capable of surviving the initial fire/explosion. These analyses led to the rectangular layout, which locates the living quarters (temporary refuge) and lifeboats at one end and the highest risks (gas compression or drilling wellbay) at the opposite end. Although the crew of an offshore production platform is close to the hazards even when off shift, the rectangular layout gives a measure of separation distance from the highest risks, which can be supplemented through the addition of fire and blast walls. Other Cullen Report recommendations addressed the need for analysis of the vulnerability of safety critical equipment or elements (SCE) such as emergency shutdown isolation valves, and evacuation, escape and rescue in the event of major incidents.

The sum total of all the above recommendations was to reduce the residual risk of the design through the use of an ISD approach. The residual risk then has to be managed by a safety management system (SMS), another inquiry recommendation. The SMS involves many of the elements of process safety, such as operating procedures, training, mechanical integrity, management of change, etc. Finally, the Cullen report recommended a Safety Case to describe and justify the design, the inherent hazards and residual risk, and the SMS. This Safety Case is now the basis for periodic audits.

[1] The term 'process safety' is not universally understood within the oil & gas exploration and production industry, although its use is increasingly widespread. Other terminology used includes technical safety, safety engineering, and asset integrity.

Subsequently, new regulations were promulgated in the UK sector of the North Sea mandating the Cullen Report recommendations. Similar regulations have been adopted in other areas, such as Norway and as far as Australia. These regulations indirectly place less reliance on the human element than was the case before the Piper Alpha disaster, although the human element is still important in delivering an effective SMS performance. Indeed, the Cullen report recognized the need for better leadership competencies for Offshore Installation Managers (OIMs), especially during emergencies. Other important human factors offshore include correctly and diligently implementing operating and maintenance procedures, permits to work, management of change, and emergency preparedness.

3. Scope of Drilling Rig Reviews

The main areas of focus for the author's reviews of D&C operations are listed in Table 1. Many of these factors are sub-elements of the safety culture, and it is important to recognize that the quality of local leadership on the rig will drive or limit the safety culture.

Table 1. Focus Areas for Review of Drilling & Completions Operations

Focus Areas for Review
Leadership
Accountability, Roles & Responsibilities
Safety Culture
Adherence to Policies, Procedures & Regulations
Technical Competency
Understanding Process Safety
Understanding and Responding to Risk
Operational Readiness
Maintaining a Sense of Vulnerability
Management of Change
Response to Anomalies and Emergencies
Planning and Timely Provision of Resources
Selection and Oversight of Contractors
Thirst for Organizational Learning
Use of Performance Indicators/Metrics

4. Good Practices

Before identifying opportunities for improvement, it is important to recognize some of the good practices that the author observed. In particular, it should be noted that a strong focus was placed on reinforcing occupational (a.k.a. personal) safety, especially on the rig floor. The majority of the personnel on the rig (i.e. rig crew) are employees of the drilling contractor. The industry has a history of serious hand injuries and dropped objects (including drill pipe), and all rig crews have had regular daily tool-box sessions and weekly safety meetings aimed at maintaining awareness of these and other occupational safety issues. Furthermore, the rig contractor and Operator leadership had high visibility on the rig floor, undoubtedly to reinforce these messages. Other good practices are listed in Table 2.

Table 2. Good Practices Observed

Operator (Oil Company)	Drilling Contractor
Bridging documents clarify roles & responsibilities, and the interface of the separate Operator / Contractor management systems	
Regular rig visits by Drilling Superintendents and other onshore D&C leadership Day and night shift Company Man (Drilling Supervisor) on rigs High visibility of Operator offshore teams on rig floors Well Operations Management Plan describes how Operator management system complies with local regulations Operator canceled poor performing rig despite schedule & cost impacts	Strong focus of offshore crews on personal safety, including: • Empowerment of workforce to stop unsafe work • Task observation practices (variations of STOP cards) • Pre-tour and weekly safety meetings • JSA prior to implementing work
Sharing lessons between Operator and rig contractors, including service providers (e.g. mud logger, cementing, etc.)	

5. Leadership

As previously stated, leadership will drive or limit the safety culture on the rig. Under normal circumstances, an oil & gas exploration company (the Operator or licensee of the exploration area) rents or leases a drilling rig from another company that owns the rig (the drilling contractor).

The drilling contractor has an onshore Rig Manager responsible for the overall operation of one or more rigs. The Rig Manager reports to a corporate executive who sets policy and standards. Most Operators have an onshore Drilling Superintendent for each contracted drilling rig, who is accountable for planning and executing all aspects of the drilling program. In addition to these shore-based leaders, there are several leaders on-site on the drilling rig, which are described in Table 3.

These responsibilities may appear confusing, but bridging documents between the Operator's and Rig Contractor's different management systems identify and clarify roles and responsibilities for leadership under various circumstances. However, it is interesting to reflect that on half of the rigs visited, at least two (and sometimes three) of the four leaders above stated in interviews that they were in charge of:

- Downhole operations, i.e. decision-making on drilling and completions work

or

- Emergencies, i.e. on-site incident commander

Table 3. On-Site Drilling Rig Leaders

Rig Contractor	OIM[2]	Most senior manager responsible for the health, welfare and safety of all personnel on board the rig and for vessel integrity issues
	Master	Licensed mariner in ultimate command (if rig is a self-propelled MODU) while under navigation, cargo operations, etc.
	Toolpusher	In charge of drilling operations (but may be rig manager on a land drilling rig)
Operator	Company Man[3]	On-site representative of the Operator who is in overall charge of the drilling and associated activities

The bridging documents are regarded as a good practice, but it is evident that several leaders did not properly understand the details within these documents. Twenty-five years ago, the Cullen Report [3] recognized the importance of leadership training and preparedness for handling emergencies. Leadership of both downhole operations and the well control emergency was also a factor in the 2010 Deepwater Horizon incident [1, 2]. It is therefore disappointing that leadership roles are not universally understood today. It is also worth noting that important HSE and well control documents were still in draft for one rig, which had already mobilized on location and was getting ready to spud.[4]

The operation of planning and drilling a well can be relatively complex and involves many activities, but there are a number of key milestones that delineate some activities from others. Drilling a well is a multi-stage project, and these key milestones are analogous to the conventional stage gates that most capital projects observe when progressing from one stage to another, such as appraisal, selection, definition, execution, and operation. Some of the key milestones involved in drilling a well are: rig mobilization, pre-spud, commencing each hole size, well testing, and pre-plug and abandon. While all of the rigs visited had a detailed well program/plan, none of them had any sort of formal "go/no go" hold points at key milestones where leadership paused to consider operational readiness of the rig and crew to progress to the next stage. Instead, it was at best an informal process and virtually seamless activity to proceed with the next task in the plan.

Two of the rigs outside European waters had operated in the North Sea at some time in the past, and therefore had a documented Safety Case. Unfortunately, since leaving the North Sea, these Safety Cases had not been maintained up to date, and were clearly not used by rig leadership, although they proudly proclaimed their existence. A Safety Case documents the design, inherent hazards, and SMS that manages residual risks. As such, its content can be a valuable tool if used proactively by leaders for self-assessment/auditing, and for training and raising awareness in new and existing employees, including leadership and supervision. The Safety Case is worthless if it sits on the shelf and collects dust.

[2] **Offshore Installation Manager**: Many offshore operators have adopted this UK offshore management model and title and applied it to their global operations irrespective of regulatory regime.

[3] **company man**: Representative of the oil company on a drilling location. Also known as drilling supervisor, rigsite leader or well site manager

[4] **spud**: The operation of drilling the first part of a new well.

Many of the leaders representing the Operator as company man (a.k.a. drilling supervisor) were on contract for a specific well or series of wells to be drilled under the duration of the rig contract. These individuals were generally unaware of the Operator's values, policies, and expectations, especially if they had not previously worked for the Operator. As a result, they approached the job in the same manner as they had done previously, and thus were not necessarily in a position to fully represent their current employer in how they conducted themselves and influenced the rig crew.

None of the onshore Drilling Superintendents or offshore rig leaders interviewed had any formal periodic management review and continuous improvement process [4]. Instead their approach and activities were generally ad-hoc and reactive in tackling the day's issues, concerns, and incidents. The adoption of a more pro-active, systematic approach to identify weaknesses before occurrence of incidents or poor performance, based around regular routine reviews of different aspects of their management system, would spur continuous improvement.

6. Competency

One of the lessons of the Deepwater Horizon incident was the emphasis placed by leadership upon occupational safety but a corresponding lack of focus on process safety. Although not mentioned by name, process safety was also one of the lessons from the Piper Alpha incident, given the number of recommendations offered regarding hazard and risk assessment, training, permit to work, and emergency response to name but a few process safety elements.

The topic of process safety was discussed during interviews with rig leaders and crew. More than one rig leader informed the author that, "We don't have processes!", "We only drill wells", or words to that effect while another leader stated, "We have processes for safety!" and then described safety meetings and other occupational safety activities and procedures. The act of drilling an oil well involves many of the elements of process safety to keep the oil in the pipe and keep the well under control at all times. However, despite some of the major incidents in the industry and the lessons that should be shared and learned, it is clear that process safety is still not well understood. This was further highlighted in discussions around barriers or layers of protection in which fundamental principles of process safety were generally not understood, despite the number of barriers being a causal factor during the temporary abandonment of the Macondo well in the Deepwater Horizon incident [1, 2].

The quality of implementation of some of the elements of process safety was variable and generally weak, especially relating to hazard evaluation, risk assessment, asset integrity, management of change (MOC), and incident investigation as discussed below. Improved knowledge and implementation of process safety represents a significant opportunity for risk reduction.

6.1 Hazard Evaluation and Risk Assessment

Hazard evaluation and risk assessment of D&C plans is necessary to determine if adequate safeguards (including barriers) exist and if further risk reduction is appropriate through the addition of other safety measures.

In general, hazard evaluation and risk assessment techniques employed on the rigs were qualitative and involved brainstorming and occasionally HAZID. None of the rig systems had been HAZOPed with the exception of one rig that had a HAZOP study conducted on its well testing system. A jack-up rig over a gas production platform did not have a comprehensive simultaneous operations (SIMOPS) plan based on a hazard evaluation of possible scenarios. A few of the rigs had risk registers, but none of them had been kept evergreen. One or two risk assessments had been conducted by the Operators, but in one case, the results of the risk assessment had not been shared with the rig leaders.

Many of the fatalities on the Piper Alpha were due to the effects of ingress of the products of combustion (smoke, carbon monoxide) into the living quarters [3]. None of the rig leaders were aware of any studies on their rig to analyze smoke and gas ingress into the engine room or living quarters, and some were unaware that gas had migrated into the engine room of the Deepwater Horizon immediately prior to the explosion and fire [1, 2]. One or two interviewees took credit for the hazardous area classification (HAC) of the rig but failed to understand that HAC only applies to the specification of electrical equipment in areas where hydrocarbons may be present *under normal operations*. The presence of vapors due to a loss of well control or other major incident is *not* normal operation.

Similarly, none of the rig leaders were aware of any studies on their rig to analyze safety critical equipment / elements (SCE) to see if it was vulnerable to the very scenarios that it was designed to mitigate, as recommended by the Cullen Report [3]. The Safety Cases, where they existed, did not address the survivability of SCE.

6.2 Asset integrity

On every rig visited, the Operator's on-site company man and onshore Drilling Superintendent were totally focused on the downhole operations. Asset integrity of the rig was left entirely to the rig contractor.

Virtually every rig visited had a significant number of overdue preventive maintenance (PM) tasks, including overdue PM tasks on what the author would deem to be SCE, such as relief valves on mud pumps or the fire and gas system. In general, the Operator's leaders were unaware of these rig integrity issues and lacked the knowledge to evaluate and challenge the rig contractor.

One rig contractor had guidance on how to identify and manage SCE within its maintenance management system, but neither of their two rigs visited had adopted this guidance and identified SCE. The concept of SCE was not understood by any of the rig and Operator leadership interviewed. Two rigs had a critical equipment list, but this was based upon required spare parts rather than safety.

Rig PM tasks were generally based on OEM's recommendations without necessarily considering the duty, required reliability, and operating history. In a few instances, where breakdowns had impacted downtime (i.e. availability), the OEM's recommendations had been changed.

The well design was undertaken by the Operator's D&C team, sometimes supplemented by support from specialist consultants. At least two of the Operators did not have a quality management process to assure the adequacy of their design work, such as requiring a senior engineer to check calculations performed by less experienced engineers.

6.3 Management of Change (MOC)

Review of rig management of change records indicated multiple changes without a formal hazard / risk assessment, and none of the rigs visited applied MOC to organizational or key personnel changes. At least one rig had late changes to its drilling plan for which MOC had not been applied and the changes had not been socialized with those who might be affected or needed to know.

6.4 Incident Investigation

Examples of incident investigation reports were reviewed on most of the rigs. None of the stated 'root causes' were true root causes, but were invariably causal factors and, at best, immediate or basic causes. If corrective actions are not based on true root causes, the actions are unlikely to be effective in preventing a recurrence of the incident or similar incidents [5].

6.5 Permit to Work

Inadequate management of a work permit was one of the causal factors identified for the Piper Alpha disaster [3]. A few of the rigs reviewed used pre-prepared generic job safety analyses (JSA) as task risk assessments to support their work permits. These JSAs were relatively comprehensive, but did not necessarily address site-specific hazards and safeguards required on the day. Other rigs did not use JSAs, while one or two rigs developed a JSA for each permit but the quality of these JSAs was poor.

One rig had adopted an electronic permit system that was managed by the ballast control supervisor who prepared and issued all work permits. Unfortunately, the supervisor was not supposed to leave the ballast control room, so the work permits were prepared in the control room and did not necessarily reflect site-specific hazards and safeguards required on the day.

6.6 Emergency Planning & Response

The author was fortunate to experience two lifeboat muster drills on one drilling rig. Despite the practice, both drills were somewhat chaotic with:

- inadequate standing room for all persons wearing life-jackets designated for the lifeboat,

- a noisy environment making it virtually impossible to hear the roll call, and
- the second spurious unplanned drill was abandoned rather haphazardly with some persons at the muster point and some still on their way.

As a first-time visitor, another rig did not show me where my designated lifeboat was located, but I deliberately walked around the rig until I found it. Later, I noticed poor housekeeping by helicopter embarking/disembarking crew members whose bags partially blocked access to my designated lifeboat.

Several rigs clearly had the potential for a Macondo scale incident on the basis of the well being drilled and its location. However, the emergency response plans did not reflect worst-case scenarios of that magnitude. At the time of the review, well containment equipment was not available for these wells like the Helix and MWCC systems developed after the Macondo oil spill for GoM wells.

6.7 Metrics

All of the rigs reviewed regularly reported traditional injury and downtime *lagging* performance indicators. However, none of the rigs had adopted any *leading* metrics [6, 7] to assist rig leadership in managing the highest risks, such as asset integrity (overdue PM tasks) and crew competency. Blowout prevention relies on rig crews recognizing a kick[5] rapidly and taking corrective action, such as closing in the well, so that another important metric could involve how long it takes crews to respond to kicks when oil, gas, or other fluids enter wellbores.

7. Lessons Learned

Each of the rigs shared information on incidents that occurred on other rigs owned by the rig contractor, and some other incidents from the drilling industry. The Operators also shared incident information with their rig contractors, although sharing of lessons between regions was variable.

All of the leadership and employees interviewed on the rigs claimed to have received information and/or attended safety meetings that shared lessons from the Deepwater Horizon incident. However, the weaknesses described above cast doubt on the effectiveness of the sharing and whether the lessons were truly learned. Most of the sharing appeared to focus on what happened rather than on the lessons and the improvement opportunities, especially those related to process safety.

One Operator had a particularly poor performing drilling rig under contract. The rig passed normal pre-qualification reviews, a technical inspection by a specialist third party used by many Operators, and an HSE review by the Operator. The technical inspection noted a few items requiring attention before the rig contract commenced. After the rig mobilized and commenced drilling, the safety performance and reliability of rig equipment were both poor. There was a significant backlog of overdue PM tasks affecting reliability. Eventually, the Operator cancelled

[5] **kick**: Entry of water, gas, oil, or other formation fluid into the wellbore during drilling. If prompt action is not taken to control the kick, or kill the well, a blowout may occur.

the contract at significant cost and delay to the drilling program. This case raises concerns about the scope of work that third party inspectors conduct, given that some of the overdue PM tasks pre-dated the original inspection and were not identified. Perhaps Operators need to tightly define the scope of third party inspections in their contracts? This also raises questions about whether third party inspectors have changed any of their practices as a result of incidents like the Deepwater Horizon, in which other equipment issues were missed.

8. Regional Variations

The strengths and weaknesses described above were not universal, and some interesting trends in safety culture were observed between the D&C teams and rigs operating in the different regions visited. While the numbers of rigs visited in each region may not be statistically significant to support these trends, the results obtained are presented in Table 4.

Table 4. Regional Trends

Region	Process Safety Understanding	Technical Program	Comments
North Sea, Australia	Good	Strongest including risk management	Due to familiarity with Safety Case regimes?
West Africa	Poor	Reasonably strong	
USA	Weakest	Reasonably strong	Weakest safety culture. Rig leadership high visibility on rig floor to compensate? Focus on personal safety only - surprising given recent SEMS regulation.
South America	Poor	Weak Planning behind schedule	Strong safety culture. Motivated to learn.
Asia Pacific (Indonesia, Malaysia, Papua New Guinea)	Poor*	Average	*except for a few ex-North Sea staff Reasonable personal safety culture, but some procedures not followed.

9. Conclusions

Preventing major accidents requires a specific focus on process safety management over and above conventional occupational safety. There is a strong emphasis on personal safety within the drilling industry, but that is not balanced by an equal focus on process safety. Process safety is not well understood, and raising awareness does not appear to have been a priority, even after lack of focus on process safety was a key lesson from one of the most recent major incidents in the industry - the Deepwater Horizon incident.

Significant weaknesses in asset integrity, especially relating to inspection, testing, and preventive maintenance of SCE, could be the failure that triggers a major incident or prevents a proper response and mitigation to a loss of well control or other cause of loss of containment. The industry needs a greater focus on integrity management, especially SCE and overdue PM tasks.

There are many opportunities to strengthen safety management systems (SMS) as recommended by the inquiry into the Piper Alpha disaster 25 years ago. Leadership appears to be too reactive and needs to adopt a more proactive approach based on formal management system review and continuous improvement.

Well, are we drilling in the right direction? Not exactly! While the industry has made great strides in developing technologies to permit exploration in ever deeper waters and reducing hand injuries and risks from dropped objects, the magnitude of the culture change necessary to advance process safety management is huge. Some rig contractors, specifically those operating in the North Sea, understand the terminology and process safety management systems, but this has not necessarily filtered down to some of the rig crews or been embraced by rigs operating in other regulatory regimes. Some other regulatory regimes, like the US OCS, are changing to mandate process safety for operating companies, but have not yet done so directly for drilling rigs.

Unless the drilling industry changes its focus on monitoring individual worker injuries while ignoring warning signs of process safety problems that have the potential to lead to major incidents, then the industry is likely destined for more disasters. The industry has the capability for directional drilling[6] to reach remote or inaccessible formations, so why not use that capability to reach a sound understanding and implementation of process safety? A workover[7] like that will certainly be drilling in the right direction!

10.References

[1] National Commission on the BP Deepwater Horizon Oil Spill and Offshore Drilling, "Deep Water, The Gulf Oil Disaster and Future of Offshore Drilling", January 2011

[2] Bartlit Jr., Fred H., "Macondo, The Gulf Oil Disaster", Chief Counsel's Report, 2011

[3] The Hon Lord Cullen, "The Public Inquiry into the Piper Alpha Disaster," HMSO, London, November 1990.

[4] CCPS, "Guidelines for Risk Based Process Safety", New York, 2007

[5] CCPS, "Guidelines for Investigating Chemical Process Incidents, 2nd ed.," New York, 2003.

[6] **directional drilling:** Intentional deviation of a wellbore from the vertical. Although wellbores are normally drilled vertically, it is sometimes necessary or advantageous to drill at an angle from the vertical to reach subsurface areas laterally remote from the point where the bit enters the earth.

[7] **workover:** Operations on a producing well to restore or increase production may be performed to stimulate the well, remove sand or wax from the wellbore, to mechanically repair the well, or for other reasons.

[6] Broadribb, Michael P., “Cheddar or Swiss,” paper presented at the AIChE 43rd Annual Loss Prevention Symposium and CCPS 24th Annual International Conference, April 2009

[7] CCPS, “Guidelines for Process Safety Metrics”, New York, 2010.

Implementation of Process Safety Key Performance Indicators in Wells and Drilling Activities at Petrobras

Felipe Augusto Farias de Carvalho
Petrobras
Avenida República do Chile, 330
Rio de Janeiro, Brazil
carvalho.felipe@petrobras.com.br

Manoela Teixeira Lopes
Petrobras
Avenida República do Chile, 330
Rio de Janeiro, Brazil
mtlopes@petrobras.com.br

Prepared for Presentation at
American Institute of Chemical Engineers
2014 Spring Meeting
10th Global Congress on Process Safety
New Orleans, LA
March 30 – April 2, 2014

Implementation of Process Safety Key Performance Indicators in Wells and Drilling Activities at Petrobras

Felipe Augusto Farias de Carvalho
Petrobras
Avenida República do Chile, 330
Rio de Janeiro, Brazil
carvalho.felipe@petrobras.com.br

Manoela Teixeira Lopes
Petrobras
Avenida República do Chile, 330
Rio de Janeiro, Brazil
mtlopes@petrobras.com.br

Keywords: Process Safety Indicators, Wells and Drilling Activities, Kick

Abstract

Monitoring and analysis of results derived from Process Safety Indicators provide inputs to the continuous improvement process in Safety Management. In fact, these results make it possible to focus efforts on the weaknesses identified and to determine potential opportunities for improvement. Furthermore, throughout the maturation time of these indicators, it is possible to verify those that are most representative for the process under analysis and the need for reformulation or exclusion of some metrics during the trial period. This paper presents a work in development at Petrobras regarding the implementation of Process Safety Indicators for the Offshore and Onshore Well Construction and Maintenance area. These metrics are discussed in relation to the strengths and weaknesses found. Additionally, analysis of the results found are presented in alignment to the relevant processes and considering the risks associated with failures of critical barriers from a Well Safety standpoint.

1. Introduction

Wells and Drilling operations involve high risk activities, so it's essential to continuously monitor the integrity of the protective barriers. An uncontrolled release of hydrocarbons for the surface may result in severe injuries to operational personnel, irreversible damage to assets and critical consequences for the environment. Therefore, an adequate process safety management is an important pillar for the business continuity. For the purposes of a continuous improvement process, measuring the performance in Safety Management may be done by means of suitable metrics that describe the system/process under analysis. These metrics may be created by identifying the existing Risk Control Systems [1] and defining the outcomes expected in case of successful performance or in a barrier failure situation.

The evaluation of Risk Control System integrity shall be done by selecting some parameters that represent the performance throughout the operational time. These metrics may be computed for each target process and may reveal some weaknesses of protective barriers before the failures occur. Thus, it is already possible to adopt measures that improve the level of safety, such as redundancy and layout modifications. Moreover, action plans may be coordinated from the moment defects are found in the system.
The outcomes of Risk Control Systems may be monitored and failures computed. It is also possible to determine defects on protective barriers, but then the losses will have already occurred. These metrics may identify an inadequate Process Safety Management, and this entails revisiting the risk control system models adopted and how they interact. These results may then be useful for reconsidering the implementation of more effective leading indicators. Proactive actions may be recommended for the normalization of the conditions before a catastrophe occurs. These metrics are called lagging indicators and their main characteristic is the fact that they monitor the outcomes of the processes under analysis, i.e., after the failures and the respective losses. Although this is a post-failure indicator, their results may be directly associated with a continuous improvement process, which is to say that these indicators would serve as an input for promoting positive changes to the process safety policy, including the reconsideration of the leading indicators model.
The implementation of leading and lagging indicators in wells and drilling activities at Petrobras may be performed taking the following steps into consideration: (i) Definition of Process Safety Anomalies, (ii) Diagnostic of Process Safety Management, (iii) Training on Process Safety competence, (iv) Identification of Process Safety metrics [2]. The first three steps encompassed all the company. The last step was accomplished for each business area separately. Otherwise, the implementation of indicators proceeds to the following stages: (a) Definition of Indicators; (b) Trial Period; (c) Full implementation of the indicators (Figure 1). A brief explanation of those indicators and the present stage of implementation at Petrobras is given on the following section.

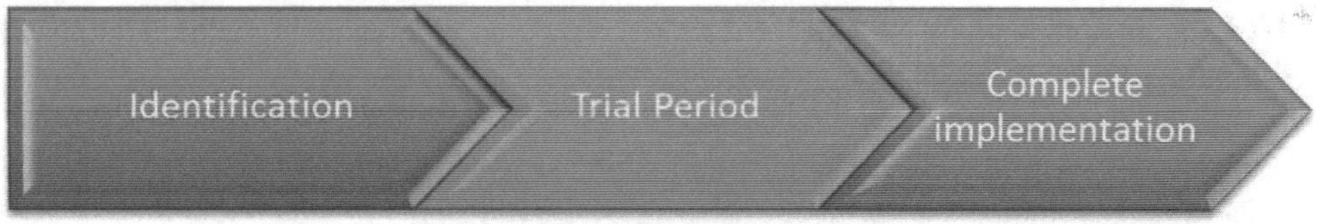

Figure 1. Process Safety Implementation stages.

2. Implementation Process of Key Performance Indicators in Wells and Drilling Activities

2.1 Identification of Risk Control Systems

The first step considered for the specification of indicators is related to the identification of Risk Control Systems. A workshop with the participation of members of areas related to these operations was held. The areas that manage onshore and offshore operations, those responsible for cementation operations and rig equipment's integrity, in addition to the Safety and Environment department, were invited for this technical event. Each area presented some potential indicators to be implemented, including the description of metrics (formula, periodicity, database, department in charge of providing the data, etc.) as well as the objectives and benefits of monitoring them. After that, the participants were divided in multidisciplinary groups to analyze each package of indicators proposed and consolidate their opinion. At the end of this event, 20 indicators were selected.

The suitability of the selected indicators was analyzed based on some publications that guide the development of indicators [1, 3 - 4]. Elementary recommendations were followed in this stage, for instance, avoiding an excessive number of indicators and eliminating indicators with no reasonable inputs or with no practical interpretation. As a result of this task, sixteen indicators were chosen to represent these activities: eleven indicators at the tactical level and five indicators at the strategic level.

The tactical metrics comprise seven leading indicators and four lagging indicators. The leading ones are related to the compliance with inspection and maintenance plans and the reliability level of well project discipline. The lagging indicators encompass measurement of pollutants leaked to sea, black outs occurred on Dynamic Positioning Rigs, recent incidents and pending issues created after regular inspections on the rig.

The metrics at the strategic level are considered as leading indicators and three of them will be presented in the next section.

2.2 Strategic Indicators

The indicators monitored at the strategic level are linked to the occurrence of a kick. The kick is an undesired inflow from the formation to the well and represents a warning of underbalance condition, i.e., the formation pressure is higher than the well pressure. In this situation, the primary control of the well was lost, i.e., the pressure imposed by the mud is insufficient to guarantee the overbalance condition [5,6].

The occurrence of the kick shall be quickly detected so that the driller and the personnel involved can act in a safe and efficient way, avoiding the release of a high volume of formation fluid to the well. In this situation, the Blowout Preventer (BOP) is actuated in order to isolate the well, blocking the upward flow to the riser (Figure 2).

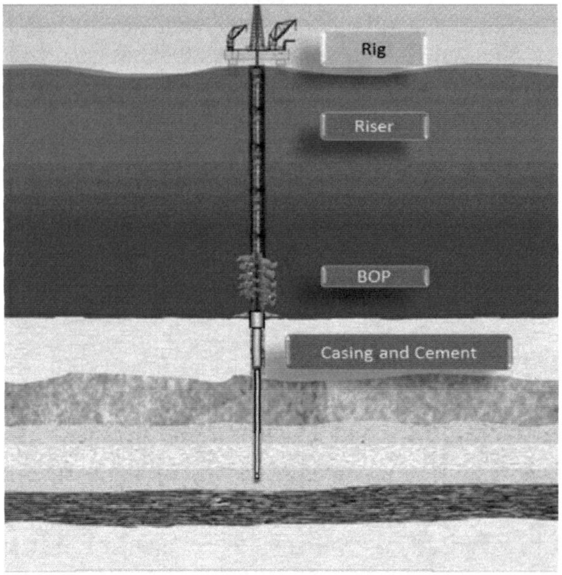

Figure 2: Illustrative Schematic of offshore operations.

Therefore, personnel involved with well control shall be periodically trained not only to avoid kicks but also to detect and control well pressure. Hence, Petrobras monitor the well control certification of onboard personnel involved in these activities. Moreover, it is important to promote kick drills onboard and monitor the volume of kicks and the time elapsed from the detection until the correct efforts to normalize well pressure. Thus, some indicators were proposed based on personnel competence to control the well. One of them is the Well Shut-In Efficiency Indicator. This metric is related to the efficiency in well shut-in during kick drills or suspected kicks. The Well Safety department established some parameters of volume and time to characterize an efficient well shut in.

Another important protective barrier is a good quality of cementation. The main objective of the primary cementation is the zonal isolation. The annulus around the casing is filled up with cement in order to restrict fluid migration between formations of different depths and to bond and support the casing [7]. Moreover, the cementation process avoids blowouts by quickly forming a seal and sealing off zones of lost circulation or thief zones [7]. For that reason, the number of cementation failures was considered and is now computed per well excluding failures in the first two phases.

The Blowout Preventer (BOP) is a piece of equipment composed by a number of rams with the role of isolating the well in case of an undesired influx (Figure 3). These rams are mechanical devices actuated by a piston mechanism in order to close against a tubular string. There are some different types of rams. The BOP represents one protective barrier for the undesired influx, so a critical failure of the BOP may make the isolation and well control via choke and kill lines unfeasible. Therefore, the Mean Time to Failure ($MTTF_{bop}$) was selected as a strategic indicator for computing BOP failures that demand its removal after installation testing.

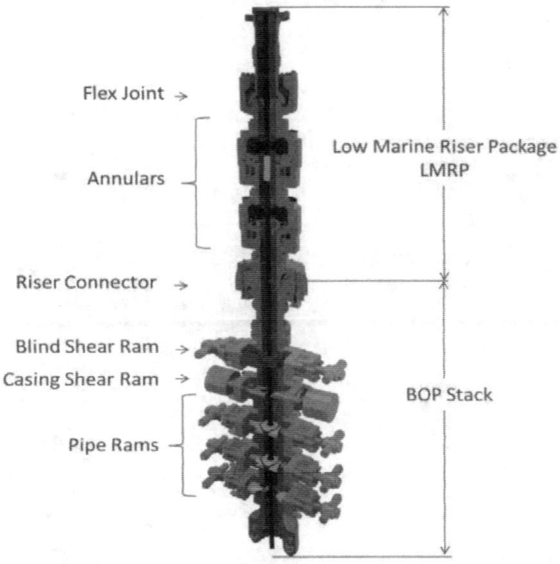

Figure 3: Typical Subsea BOP

In the same way of the metrics previous discussed, the fifth strategic indicator selected is directly associated with kick occurrence. The number of kick events is computed on a per-well basis encompassing drilling, completion and workover operations. Although kick occurrence is an abnormal well condition, it is still possible to control it. Therefore, this indicator is considered a leading one in relation to the blowout event. If actions are taken to reduce the number of kicks, the probability of a blowout occurrence declines.

2.3 Trial Period Stage

After the first stage of identifying risk control systems and selecting suitable metrics, the indicators started being used. At this moment, they are assessed as a representative model of risk control systems integrity (leading indicators) or as a warning of barriers weaknesses (lagging indicators) for the process under analysis. Hence, it is still possible to reformulate the indicator, to establish a suitable periodicity for consolidating the data or, for instance, excluding some unreasonable indicators.
The trial period is helpful for defining indicator's goals.
The following section presents more details about such strategic indicator.

3. Deepwater Drilling and Well Integrity Safety Indicators.

Often, a major challenge is that there is not enough data to support a basic set of reliable and valid safety indicators. Therefore, a broad perspective is needed when developing and analyzing indicators. In addition, the safety indicators should reflect the phases of deepwater drilling (and drilling in general). These phases are the well planning phase and the drilling phase, consisting of drilling, running casing, cementing, circulation, fluid displacement and clean-up, and completion [8].

Figure 4 gives an overview of the main lifecycle phases of the well and aspects related to undesired incidents. It also presents relevant barriers of deepwater drilling [8].

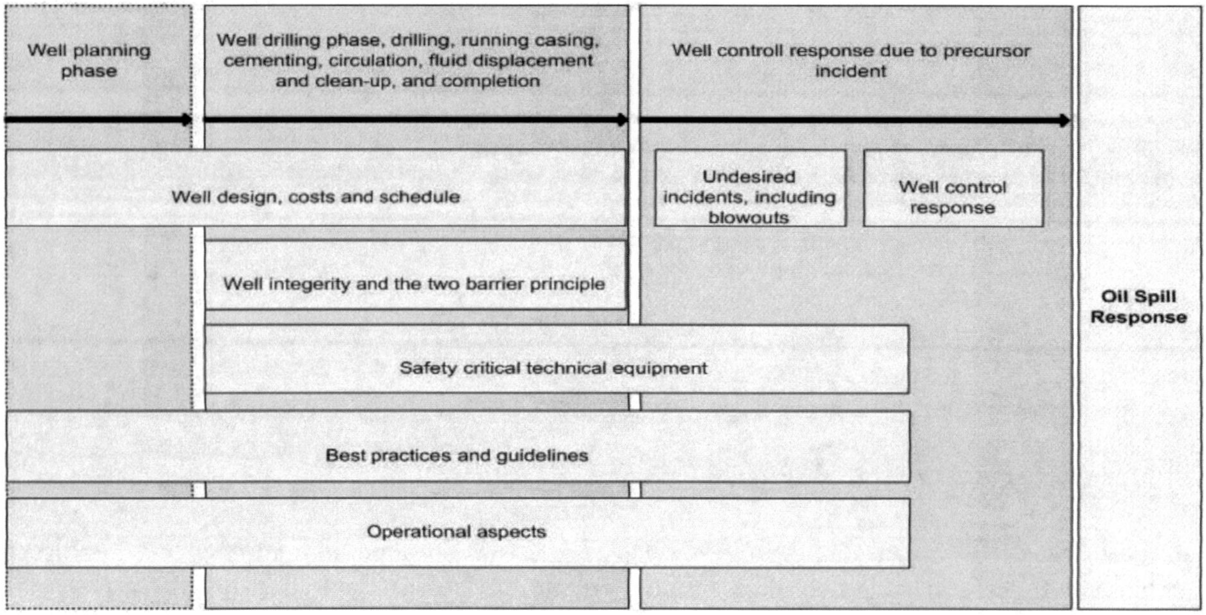

Figure 4: The life cycle phases of a well (excluding production)

4. "Kick Package"

Kick Indicator - Ikick

A kick is instability in the well as a result of the well taking in gas, oil or water, and may lead to a blowout. A blowout in turn is defined as an unintentional flow of formation fluid from the well to surroundings or between the formation layers after the defined technical barriers, and the operation of those, have failed [8].

According to a study performed by MMS, 39 blowouts occurred in the GoM in the 15-year period from 1992 to 2006, and this would mean an average ranging between 2 and 3 per year. Such a number would be too small to present itself as a useful indicator.
That said, an immediate precursor to a blowout can be considered, namely a well kick (or well-control incident, as it is also called). These happen more often, and it is widely recognised that by reducing the number of kicks, the risk of blowout is also reduced. For any particular well, there might be too small a number of kicks so as to serve as a useful indicator, but number per company per year is something that could be usefully computed by companies aiming to drive this number downwards [10].

The GoM frequency of deepwater kicks is high. The overall frequency of kicks is approximately 2.7 times higher in the US GoM deepwater wells than the overall Norwegian Continental Shelf (NCS) experience. That said, the NCS kicks in deep wells, and especially HPHT wells, have occurred frequently [8].

Purpose of Indicator:

The company has therefore defined the Ikick – Indicador de Kicks (Kick Indicator) as the main indicator, aiming to monitor the occurrence of Kicks per well.

Calculation

This indicator is calculated as follows:

$$\textbf{Ikick} = \textbf{(NAK / NIPO) * 1000 [Eq. 1]}$$

Where: NAK – Number of Kicks and NIPO – Number of wells (Drilling, Completion and Workover Operations).

This indicator is measured on a monthly basis, and is reviewed by the executive managements every quarter.

Results

According with Jon [8], NCS experienced a frequency (annual average 2003-08) of 16,2 Well kicks/loss of well control in this period.

In 2013 Petrobras experienced a lower kick frequency in Brazilian Continental Shelf for development wells in comparison to all oil and gas production installations and mobile drilling units which have operated on the Norwegian Continental Shelf (NCS) in 2003-2008.
Also in Brazil, HPHT wells have higher reportable kick incident rates compared to non-HPHT wells.

Well Control

Well control procedures are established to safely prevent or handle kicks and reestablish primary well control. The number of kicks and blowouts are relevant indicators, but there is a need for developing a set of deepwater drilling indicators for precursor incidents leading up to those kicks and blowouts [8].

Well Shut-In Efficiency Indicator - IEFP

The blowout risk has another potential indicator, and this indicator became apparent during the inquiries following the Macondo accident. Blowout prevention depends mainly on drillers recognising kicks as quickly as possible after their occurrence, and taking corrective action, such as closing the well. The night the Macondo blowout happened, drillers took around 40 minutes to recognise the occurrence of a kick, and by then it was too late. About a month prior to that, the Deepwater Horizon had another kick; this one took 33 minutes to be recognised. An analysis that followed indicated that it should have been recognised much earlier. That said, it may be easy to imagine an indicator that is based on response time to kicks, and such an indicator would be relevant for a company or for the entire industry, perhaps even at the level of individual wells.
All the data are recorded in an automatic fashion, so, as previously mentioned, this would be an indicator with a significant amount of robustness. Interestingly, companies performed tests of response time. Such tests involved a simulated kick to check how long it took workers to recognise and respond to the changed circumstances. This could also serve as the foundation for a rig-level indicator that showed how well blowout risk is being managed [10].

Purpose of Indicator:

With this, Petrobras has established criteria for efficient well shut-in, with defined standards and time. Therefore, the IEFP – Índice de Eficiência de Fechamento de Poço (Well Shut-In Efficiency Indicator) was created as an indicator, with the goal of assessing well shut-in efficiency in case of a kick or a kick drill.

Calculation

This indicator is calculated as follows:

IEFP = (NFTP / NEK)*100 [Eq. 2]

Where: NFTP – Number of Shut-in Activities Using the Correct Procedure, and Within the Expected Time and NEK – Number of Events (kick drills + suspected kicks) (Drilling, Completion and Workover Operations).

Objective criteria shall be defined to qualify a correct shut-in, and non-compliance with any of such criteria shall render the shut-in incorrect. Examples: 1) Whether the shut-in procedure is according to the standard 2) Whether the shut-in time is coherent with the mean times seen during simulated kicks. 3) Whether the kick volume is below the expected for the Project.

This indicator is measured on a monthly basis, and is reviewed by the executive managements every quarter.

Results

Due to the short period in which this metric is being used, there were no conclusive results until the closing date of this paper.

Cement Failure Indicator - ICIM

Another indicator of blowout risk that might be useful is the number of cement failures. The Macondo blowout started with a cementing failure that was not recognised. In addition to that, two previous cementing failures had occurred higher up the well. The MMS study mentioned earlier found that 18 of the 39 blowouts in the 15-year period under consideration had been initiated by cementing failures. So, reducing the frequency of cement failure would be desirable not only from a commercial standpoint, but also from a safety point of view. Number of cement failures is an indicator that the regulator should consider tracking [10].

Purpose of Indicator:

Aiming to identify and monitor the frequency of cement failures, the ICim – Indicador de Falhas de Cimentação (Cement Failure Indicator) was created.

Calculation

This indicator is calculated as follows:

$$\mathbf{ICim = (NFCim / NCim)*100 \quad [Eq. 3]}$$

Where: NFCim – Number of cement failures and NCim – Number of accomplished cementation

This indicator is measured on a monthly basis, and is reviewed by the executive managements every quarter.

Results

In the GoM, there were 20 incidents from 1973 to 1995 related to well kicks after cementing surface casing. Another 13 similar incidents have occurred since 1995, with the most serious consequences being gas broaching to the surface, cratering, well loss, and rig and platform destruction by fire. Annular flow related to cementing surface casing has been identified as one of the most frequent causes of loss of control incidents in the GoM [8].

In comparison with this scenario, Petrobras experienced a smaller number of cement failures in 2013.

Mean Time to Failure BOP - MTTFbop

An important part of the Performance Standard related to well integrity is the maintenance and inspection of the BOP.
At the Deepwater Horizon rig, the BOP did not isolate the well before and after the explosions. The BOP may have been faulty before the blowout or it may have been damaged due to the accident. According to BP, several maintenance jobs of the BOP were overdue, and leaks from the hydraulic control system had been discovered at the time of the accident. The BOP on the Deepwater Horizon was not recertified in accordance with federal regulations because the certification process would require full disassembly and more than 90 days of downtime [8].

This metric evaluates the technical condition of this safety critical equipment.
The safety critical equipment is comprised of mechanical defences or barriers that should be positioned to prevent major accidents. Monitoring the effectiveness of such equipment is a crucial activity for managing major hazard risk [2,10].

Purpose of Indicator:
With the purpose of evaluating the mean time for critical period test failure, the MTTFbop – Mean Time to Failure BOP was created.

Calculation

This indicator is calculated as follows:

MTTFbop = (Σ TEXPbop (stand by) / Σ Nfalhas) [Eq. 4]

Where: Σ TEXPbop (stand-by) – Sum of BOP exposure days and Σ Nfalhas – Sum of all failures.

Results

A better performance was observed in 2013 in comparison with 2012; the MTTFbop increases almost 40%. The value obtained in the last two years is higher than the results presented in JIP Study related to the wells drilled in the US Gulf of Mexico in 2004 to 2006 for Subsea BOPs. In this aforementioned reference, the MTTF is approximately 338, but it includes all the failures, not only those related to BOP removal demand [11].

4. Conclusion

There is a common understanding among the regulatory authorities in Norway, UK, Canada and the U.S. that "kicks" are precursor incidents and should be avoided. The probability of kicks depends on geological conditions, but kicks can be prevented by proper well planning, design, and performance monitoring [8].

Petrobras understands the importance of identifying frequently occurring precursors to kicks, and seeks to drive down their number.

Since well incidents are the most significant contributors to kicks, by driving down the number of such incidents, we reduce the risk of kicks and hence the risk of blowout [2,10].

Safety indicators are only worth developing if they are used to drive improvement, that is, if the organization focuses attention on them in such as way as to make a difference. In order to reach this goal, Petrobras works with the continuous and visible support from its executive management through the provision of resources, training/awareness and commitment [2].

The definition of an integrated set of leading and lagging performance indicators, which are periodically updated, will provide Petrobras with the possibility of achieving continuous improvement in process safety management.

5. References

[1] *Developing process safety indicators: A step-by-step guide for chemical and major hazard industries*, HSE publication.

[2] CARVALHO, F. A. F., LOPES, M. T., "Process Safety Key Performance Indicators in Wells and Drilling Activities at Petrobras", *9th Global Congress on Process Safety*, San Antonio, Texas, USA, April 28 – May 1, 2013.

[3] ANSI/API Recommended Practice 754, Process Safety Performance Indicators for the Refining and Petrochemical Industries.

[4] CEFIC, "Guidance on Process Safety Performance Indicators", 2011.

[5] SANTOS, O. L. A., *Segurança de Poço na Perfuração*, Petrobras Training Notes.

[6] SANTOS, O. L. A, 2013, *Segurança de Poço na Perfuração*, 1 ed. São Paulo, Blucher.

[7] Petrowiki.spe.org. *Cementing Operations*, 2014. Available from: http://petrowiki.spe.org/Cementing_operations.

[8] *Looking Back and Forward: Could Safety Indicators Have Given Early Warnings about the Deepwater Horizon Accident?*, Deepwater Horizon Study Group.

[9] HOLLAND, P., SKALLE, P., "Deepwater Kicks and BOP Performance, Unrestricted version," Minerals.

[10] HOPKINS, A., "Safety Indicators for Offshore Drilling: A working paper for the CSB inquiry into the Macondo blowout", 2011.

[11] SAFFLER, J., GALLANDER, F., "Just How Reliable is your BOP today? Results from a JIP, US GOM 2004-2006", IADC/SPE 128941.

Are we really learning from major accidents in Drilling operations?

Claudio Castaneda
EQUION Energia Ltd
Carrera 9A #99-02 9th Floor
claudio.castaneda@equion-energia.com

Prepared for Presentation at
American Institute of Chemical Engineers
2014 Spring Meeting
10th Global Congress on Process Safety
New Orleans, LA
March 30 – April 2, 2014

UNPUBLISHED

Are we really learning from major accidents in Drilling operations?

Claudio Castaneda
EQUION Energia Ltd
Carrera 9A #99-02 9th Floor
claudio.castaneda@equion-energia.com

Keywords: learning, major incidents, process safety, engineered solutions.

Abstract

Recently, we past the 25th anniversary of the Piper alpha disaster and also three years after Macondo incident where lives were lost, significant impact to the environment not to mention economic and reputation effects.
In the onshore drilling operations, people tend to think that these major problems are only associated to the offshore drilling, where in the middle of the sea no one can help except you. WRONG!!! Well control whether at the sea or land can take people´s life and cause companies to be out of business in an instant.

Standardization organizations like API, NORSOK among others have made efforts to bring process safety to the top priority issuing several documents and rules that help operators and drilling contractors to find a systematic way to enhance HSE while adopting process safety principles by enforcing mainly hard barriers -equipment, process, to "soft" barriers people competence and performance as a way to "measure" how industry is doing safer drilling operations.

EQUION taking advantage of its legacy - Former BP Colombia - has implemented a process safety strategy for its drilling, completions and interventions operations in the Casanare foothills jointly with major drilling and service companies to address major accident risks including: well control, pumping fluids, managing heavy loads and losing well integrity which have created in Colombia a new culture towards PS and a new way to conduct business in a responsible way although our regulations are still far away from a demanding point and thus most of the operators are really behind and experiencing serious incidents related mainly to people being injured and spills affecting the environment.

As mentioned above, the PS strategy established a number of requirements and associated KPIs for each one of the four pillars: Equipment, processes, people and the way performance help to achieve a true culture towards process safety in our drilling operations. Although big steps toward safe operations are implemented, a long walk to avoid major incidents is still to be run as well as confidence in the oil industry performance.

1. Introduction

In general, many offshore and onshore drilling and completions (D&C) operations today have the potential for a major incident related to loss of well control or fire/explosions derived from simultaneous operations with consequences such as the ones derived from major incidents including loss of lives, environmental impact and economic losses. These major hazards are related not only to human factors such as safety culture or competences but safety critical equipment failure, emergency response and weaknesses on well design and barriers [1]. Refer to Figure 1.

During the last years, operators and drilling contractors have enforced compliance of regulations, built stronger SEMS systems, improve performance indicators and strengthen HSE safety cases, built mainly in the offshore operations. Some of the root causes leading to the Macondo disaster were related to engineering design failures that appears not to be widely disseminated i.e zonal isolation, cement design for high temperature, centralization of casing-liner strings, acceptance criteria for negative tests and so forth.

If we extrapolate learnings from Piper Alpha and Macondo disasters in the onshore drilling and developing countries, our companies have focused most of their efforts in improving people competence, enhancing equipment integrity & reliability, responding to emergencies at a huge scale but not properly addressing engineering solutions to new well challenges when pushing the design limits in terms of higher temperatures, pressures and deeper wells in remote areas.

Many of the lessons derived from major incidents are still valid in areas of low norms enforcement due to weak regulations especially in the onshore business of developing countries. Our company have implemented a process safety involving engineering, asset integrity, major hazard preparedness, competency of people to capitalize learning from disasters by adopting risk based process safety principles from CCPS customized to onshore drilling along with IADC – API related standards setting expectations for key elements and performance indicators.

As an industry, we are not to forget or ignore these lessons to avoid repeat some of these mistakes or learn in the hard way. We cannot afford to have another Macondo or Piper Alpha anywhere in the world!!

2. Key learnings from Piper Alpha and Macondo

Early in july 1988, a major disaster in the North sea took the lives of 165 persons on board of Piper Alpha platform. This incident became the worst offshore oil industry disaster and over 100 recommendations from the Cullen report have been adopted included mandatory safety cases for the UK North Sea as well as safety enforcement from government agencies.[3] IADC developed and refined HSE case guidelines for both offshore and land drilling. The European Union recently adopted an offshore safety directive using UK system that will be expanded to the Mediterranean, western Atlantic and some areas of the Black sea.

The Macondo (Deepwater Horizon) incident "On the evening of April 20, 2010, control of the well was lost, allowing hydrocarbons to enter the drilling riser and reach the Deepwater Horizon, resulting in explosions and subsequent fires. The fires continued to burn for approximately 36 hours. The rig sank on April 22, 2010. Several efforts were made to seal the well, until it was permanently plugged with cement and "killed" on September 19, 2010.

Investigation report [2] from commission to the US President concluded among others that:

- The explosive loss of the Macondo well could have been prevented.
- The immediate causes of the Macondo well blowout can be traced to a series of identifiable mistakes made by involved companies** that reveal such systematic failures in risk management that they place in doubt the safety culture of the entire industry.
- Deepwater energy exploration and production, particularly at the frontiers of experience, involve risks for which neither industry nor government has been adequately prepared, but for which they can and must be prepared in the future.

Further reports on technical causes of Macondo incident related four areas with eight barriers that were breached to this incident:

- Well Integrity was not established or failed (*Annulus cement barrier and shoe track barriers did not isolate hydrocarbons)*.
- Hydrocarbons entered the well undetected and well control was lost (*Negative pressure test was accepted although well integrity had not been established, influx was not recognized until hydrocarbons were in riser and well control actions failed to regain control of well*)
- Hydrocarbons ignited on the Deepwater Horizon (*Diversion to mud gas separator resulted in gas venting onto rig and fire & gas system did not prevent hydrocarbon ignition*), and
- Blowout preventer did not seal the well (*Blowout preventer – BOP- emergency mode did not seal well*).

These incident root causes forced Operator a thorough review of design /operating practices and management systems, oversight and assurance of main contractors across of their operations worldwide. Other main operators acted on their own to capitalize learnings from Macondo by enforcing design, operating practices and being prepare to major incident response.

Macondo incident derived in the enforcement of regulations, rules and standards to drill offshore including a moratorium to new deep water drilling until new demanding rules on safety and environmental management systems (SEMS) were implemented by November 2011. The minerals management service MMS was replaced by the Bureau of Ocean Energy Management, Regulations and Enforcement (BOEMRE) focused on the Outer Continental Shelf (Mainly directed to Gulf of Mexico GoM activities) [4] . At the same time, a new agency - the Bureau of Safety and Environmental Enforcement (BSEE) –was created to regulate oil, gas, and sulphur exploration, development, and production operations on the OCS. Figure 1 illustrates well control incidents trend in OCS areas bot SINTEF´s blowout databases also shown non decreasing trends.

Figure 1. OCS Loss of Well Control: CY 2007– 2013

LOSS OF WELL CONTROL ***	2007		2008		2009		2010		2011		2012		2013	
	GOM	PAC	GOM	PAC	GOM	PAC	GOM	PAC	GOM	PAC	GOM	PAC	GOM	PAC
Flow Underground	1	0	1	0	0	0	0	0	0	0	0	0	1	0
Flow Surface	3	0	3	0	2	0	1	0	2	0	4	0	4	0
Diverter Flow	0	0	1	0	0	0	0	0	0	0	0	0	3	0
Surface Equipment Failure	3	0	3	0	4	0	3	0	1	0	0	0	1	0
LOSS OF WELL CONTROL (TOTALS)	7	0	8	0	6	0	4	0	3	0	4	0	9	0
COMBINED TOTAL FOR THE YEAR	7		8		6		4		3		4		9	

SOURCE: BSEE Database as of 14-Jan-2014

United States recent incidents in the US like refinery explosions recommended a lot of work in process safety management areas [4]. Some operators have wrongly interpreted notorious improvement in personal safety as improvement is major risk control or improvements in process safety. Other examples of offshore incident happened in 2009 in Australia with the Montara blowout that having implemented a safety case for their installations, the well risks were not completely addressed. West Delta Block incident in GoM in 2012 caused an explosion, fire and Fatalities that *remind us that those incidents are avoidable*.

Clear differentiations although cooperation is required to enhance process safety from personal safety as the consequence of materialization of PS risks are demonstrated to be huge. Not only offshore but onshore drilling risks related to well control remain constant and thus action is required from engineering, planning, equipment reliability, preparedness to response emergencies, competency of personnel and more challenging standards and regulations among of awareness towards major risks and consequences to the environment, people and economics.

To be fair, *Safety cases introduced since Piper Alpha as well as implementation of safety and environmental management plans then turned into systems have had a significant impact on safer operations in offshore activities*. For onshore activities mainly related to production facilities a number of rules have been implemented to enhance Process Safety Management (PSM) standards from OSHA, CCPS and other organizations which do not have marked differences but address almost same programs elements as shown in figure 2 below.

OSHA 1910
1. Employee Participation
2. Process Safety Information
3. **Process Hazards Analysis**
4. Operating Procedures
5. Training
6. Contractors
7. Prestartup Safety Review
8. Mechanical Integrity
9. Hot Work
10. Management of Change
11. Incident Investigation
12. Emergency Planning and Response
13. Compliance Audits
14. Trade Secrets

API RP 75 (*)
1. General – Principles & Scope
2. Safety & Environmental Information.
3. **Hazards Analysis**
4. Management of Change
5. Operating Procedures
6. Safe Work Practices
7. Training
8. Mechanical Integrity
9. Pre Startup Review
10. Emergency Response & Control
11 . Investigation of Accidents
12. Auditing Program
13. Records & Documentation

CCPS-RBPS
1. Process Safety Culture
2. Compliance with standards
3. Competence
4. Workforce Involment
5. Stakeholder Outreach
6. Knowledge Management
7. **Hazard Identification / Risk**
8. Operating Procedures
9. Safe Work Practices
10. Asset Integrity / Reliability
11. Contractor Management
12. Training / Performance
13. Management of Change
14. Operational Readiness
15. Conduct of Operations
16. Emergency Management
17. Incident Investigation
18. Measurement and Metrics
19. Auditing
20. Management Review

Figure 2 – Process safety management standards. *For offshore operations

Commitment towards learning

No doubt that management commitment and leaders driving process safety culture at rigs, make a difference in safety culture performance of their operations for good or for bad. Therefore, leadership is key to enforce prevention, mitigation and reaction upon problems but consistently working those major risks from early project stages. At the site, company men, rig managers, Offshore installation managers among others are the ones responsible for their work forces with a clear role of bringing them home back as safe as they reach the rig sites.

Most of big Operators took very serious process safety risks at the rig floor like having certified and function tested blow out preventers, well control competencies and drills, thoroughly inspecting rig structures for objects that can drop, operation of safety systems for top drives, review all high pressure systems and pressure relief devices, safety meetings, permits to work processes among others that would save lives should an emergency occur. Some others adopted too systematic approaches to make wells safer by designing for control and securing containment.

In my experience, most of the standards addressing weaknesses identified in major incidents are implemented in offshore platforms but engineering aspects like minimum requirements for zonal isolation, ruling of positive and draw down test acceptance, mechanical barriers requirements, are delayed.

Visits and commissioning of rigs in onshore drilling activities in Latin America demonstrated willingness to learn, motivation to work safely but lack of consistency in the implementation of process to ensure asset integrity, lack of spare critical spare parts, homemade tools, reluctance to report / dig deeper in operational incidents looking for real root causes and poor management of change barely touching people or process and some control for equipment repairs or modifications. When aspects like BOP certification, structure certification, OEM comes to the discussion; a long way is still to be walked.

Managed pressure drilling (MPD) [5] is recognized as a key enabling technology for drilling challenging wells. The ability to monitor and control volume in increments smaller than one barrel can help to resolve a wide range of downhole problems, such as kicks, losses, wellbore stability and drilling through narrow mud windows which can be used both offshore and onshore drilling are industry steps to get effective risk reduction.

Developing a process safety culture in Drilling activities

Our company although not operating offshore projects now (We drilled then P&A successfully an offshore well to 13,000 ft TD with 150 ft water column in the Caribbean sea in 3Q´12), is implementing process safety in drilling activities extracting lessons from other operations and focusing in key aspects of process safety standards as described in figure 3.

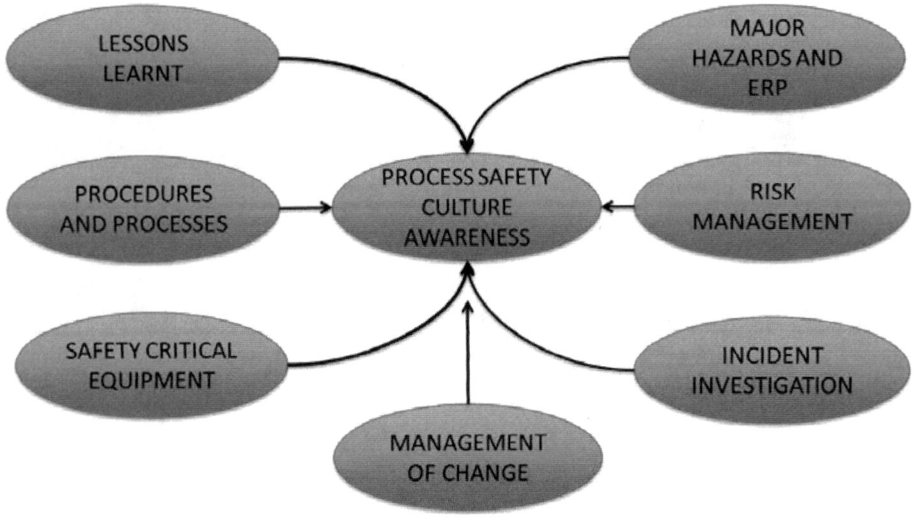

Figure 3. Process safety model implemented for onshore drilling

What´s next??

Two years after Operating companies in the OCS have to be in compliance with SEMS rule (November 15th, 2011) according to the BOEMRE, although a lot of progress in major oil companies is evident, incidents like Deepwater Horizon (Macondo well) are so impacting that not only SEMS requirements are to be met but a lot of engineering/technical practices are still to be developed as well as key performance indicators and management process into Operators to give confidence to the public that our drilling operations are safe enough to reduce the likelihood of such type of major incidents here in referred to.

The cooperative efforts among big Operators have clearly demonstrated that jointly like in the design and construction of a 15K rated 100,000 bopd capacity capping stack emergency response systems that can be installed to the wellhead, BOP or top of the LMRP, which can be available to the site in a matter of few days. Early prevention systems and reaction to well influx like MPD systems increase the level of confidence and make our operations safer for people, environment and conduct of business.

Development and deployment of next generation well control involving global standards for well designs, improve well control training-certification programs, automation of kick detection, reliable BOP - associated equipment and emergency response enforcing verification of well barriers effectiveness will enhance oil industry confidence.

Requirements for relief wells, simultaneous well emergencies in offshore areas, adoption of engineering minimum requirements for zonal isolation and drawdown tests are still areas of focus to the industry. Oil spill response to big scale is an area of further development for deployment in remote areas. The International Association of Oil & Gas Producers (OGP) created a Global Industry Response Group- GIRP which has set recommendations on well incident prevention, intervention and response capabilities after these major incidents working together with national and international regulators and industry associations that will reduce the likelihood and consequences of such major incidents.

BOEMRE 30 CFR part §250.1920 states auditing requirements for SEMS programs and results from these audits to SEMS programs under new rules are to be spread among Operators and drilling contractors – service companies to help focus on industry efforts to improve process safety in drilling operations

Conclusions

We do learn from incidents but our well barriers are weak!! Offshore industry learnt many lessons from major incidents and a number of efforts are ongoing to increase confidence in sound development of resources as result of standards and enforcement but a clear commitment to operate. Major accident hazards became a top priority.

Onshore oil and gas industry is fairly behind processes implemented for offshore activities, but no clear plans or enforcement is conducted.

Stronger or more prescriptive regulations are to be developed or adopted by local regulators including rules enforcement of drilling and production activities. Operators MUST commit their operations to conform to sound practices to preserve, protect, and develop mineral resources by adopting OGP core areas recommendations.

Large Operators designed and deployed 100 Kbbls capacity containment systems that within days can be available at remote offshore locations. Technology and lessons from other parties like Navy and airlines are also present to increase kick detection, management and equipment reliability by increasing automation.

Today, **risk for loss of well control is still present in our industry** with probability to repeat costly incidents especially in moderate to low size lease areas among of increased well challenges and complexity.

References

[1] Broadribb, Michael P., "WELL, WELL!! Are we drilling in the right Direction?. Latin America process safety Congress. Cartagena. August, 2013

[2] National Commission on the BP Deepwater Horizon Oil Spill and Offshore Drilling, "Deep water, The Gulf Oil Disaster and Future of Offshore Drilling", January 2011.

[3] The Honorable Lord Cullen, "The Public Inquiry into the Piper Alpha Disaster", HMSO, London, November 1990.

[4] US Chemical Safety Board Investigation over: At the Time of 2010 Gulf Blowout, Transocean, BP, Industry Associations, and Government Offshore Regulators Had Not Effectively Learned Critical Lessons from 2005 BP Texas City Refinery Explosion in Implementing Safety Performance Indicators.

[5] IADC, International Association of Drilling Contractors. Drilling Contractor magazines. July-August; September October. 2013.

352789 A Summary of the Solids Founds at the Modular Caustic-Side Solvent Extraction Unit Process

Abstract:

Since March 2008, the Actinide Removal Process-Modular Caustic-Side Solvent Extraction Unit (ARP-MCU) facility has successfully processed more than three million gallons of supernate. The process has experienced a few outages arising from plugged coalescers. Recently, the facility began an outage to perform maintenance activities aimed at extending the original life cycle for the facility in light of the future deployment of the Next Generation Solvent (NGS).

The outage included cleaning and maintenance of the centrifugal contactors. Inspection of the extraction centrifuges revealed the presence of accumulated solids in the bottom channels of Extraction Contactor SEP-401. SEP-401, the first centrifuge the supernate enters the solvent extraction process, is where radioactive supernate from ARP meets solvent scrubbed with 0.05 M nitric acid. The aqueous carry over from the solvent mixes with the supernate possibly resulting in a lowering of its pH to the point of solid precipitation.

Any large variance in the concentration of the aluminum and silicon concentrations, elements known to readily precipitate throughout the tank farm, may indicate precipitation driven by a mild pH swing. A similar deposit discovery (sodium aluminosilicate) was also found on September 2010 at the Salt Solution Feed Tank.

This poster presents a summary of the solids found throughout the different components that are part of the MCU at the Savannah River Site (SRS)

352853 Fighting Scales in Oil and Gas Production Systems – a Challenging Flow Assurance Case Study

Abstract:

Maturing oilfields has not just caused problems of large water handling but also demands novel flow assurance solutions which are practical and cost-effective. One of the major problems experienced in the oil production industry is the formation of mineral scale. Scales can form from the formation water as it undergoes changes in pressure and temperature, and tend to deposit on surfaces. Its formation causes flow reduction or even blocking of pipes, valves and other process equipment.

The main implications of the problem are:

- Economic; due to downtime equipment replacement, reduced or lost production
- Safety; Equipment failures due to under-deposit corrosion can cost lives.

Summarized, the impact of scales in Oil & Gas surface equipment can be serious and challenges in removal, prediction and prevention are numerous.

An approach was developed for utilizing prediction-based knowledge on the scaling problem to formulate a workflow for flow assurance monitoring. The workflow is also intended to aid in arriving at practical and cost-effective mitigation strategy for problems related to scaling in surface equipment. The following are the brief steps involved in the workflow:

- Identify & define the key Issues (such as production shortfalls, pressure anomalies, etc.)
- Define objectives and specify boundaries
- Model process (multiphase network model)
- Analyze Results and employ a variety of techniques to pin down the causes
- Conduct a structured assessment to determine/evaluate solutions
- Develop an action plan / process redesign for implementing the solutions

The root causes were confirmed as calcite scale deposition in flow lines. A detailed assessment was also performed to ascertain thermodynamic envelopes for solids deposition and identify specific hot spots related to pressure, temperature in the network to ensure trouble-free production.

The workflow suggested was valuable for troubleshooting bottlenecks of existing facilities & addressing production shortfalls. The lessons learned in this endeavor were also used to make better decisions for designing new projects.

352615 Computational Study of Oil-Water Wetting Regimes Established in Horizontal Flows: Wettability Maps and Corrosion

Abstract:

Wettability effects in pipelines carrying oil-water phases at varying proportions are crucial in undermining the net effects of corrosion in the system. Prediction of phase distribution, especially the information whether oil or water wets the pipe wall is an important factor, which can affect the corrosion mitigation strategy and increase confidence in the integrity of the pipeline. There are large knowledge gaps in this area of research and consequently only empirical criteria are used in the industry.

In the present work, detailed computational study involving, a turbulent oil-water multiphase flow system in a horizontal carbon steel pipeline, is carried out to predict different water wetting regimes as well as corrosion kinetics of emanating from the fluid-metal interaction. Computational Fluid Dynamics simulations include Eulerian multiphase models with appropriate sub-models for interfacial forces, turbulent interaction and population balance equations. Population balance techniques identify and follow the dispersed (water) phase dynamics such as droplet breakage and agglomeration in a detailed fashion. In addition, concentration equation for ferrous ion ($Fe2+$) is modeled to capture corrosion kinetics. The current study focuses on flow pattern computations for water cut ranging from 0 - 20%.

Our simulations captured various water wetting regimes inclusive of intermittent, fully wetting, dispersed and semi-dispersed water-in-oil behavior. Results were in good agreement with the experimental observation. Simulations clearly predicted that for given superficial water velocity, elevating the superficial oil velocity leads to transition of regimes from stable water wetting to intermittent water wetting and subsequently stable oil wetting scenario. Phasic fraction and velocity distribution at different flow cross-section reveal the flow modulations as a function of the turbulence inherent to the two-phase system. Corrosion computations clearly indicate a substantial generation of Fe2+ ion under stable water wetting regime while the concentration and corrosion levels deteriorate with intermittent water wetting, an effect well characterized in flow loop tests.

352338 Modeling of Conductive Antifouling Paints for Marine Applications

Abstract:

Protection from fouling organisms (biofouling) is a key function of coatings on marine hydrokinetic (MHK) devices. Accumulation of biomass on devices risks performance degradation and possible failure. At the present there are no proven reliable options for prevention of fouling in the marine environment with applications to match device lifetimes expected to be on the order of 10-20 years. Removal, cleaning, and re-application of short-lived antifouling paints is an expensive prospect. Long term solutions for antifouling protection need to be identified and tested for reliability. Electrified conductive paints have been shown to be effective in the prevention of marine biofouling and may be a viable solution for long term antifouling performance[1] . The present work seeks to further the understanding of the mechanism of action and to develop the engineering knowledge needed to enable application.

The mechanism of action has been proposed variously to be direct electron transport[1], electromagnetic field repulsion[2], and electrochemical reaction[3, 4]. Evidence suggests that electrochemical oxidation of chloride and bromide present in seawater produces strongly oxidizing agents which sanitize the surface being protected. These oxidizing agents are short lived in the environment with end states being the original chloride and bromide. Comparison of the experimentally determined minimum potential needed for biofouling protection with Tafel plots in sterilized and newly acquired seawater support this hypothesis, as does cyclic voltammetry.

Chloride oxidation at graphitic anodes has been studied in highly saline water at high over-potential for the chloralkali process. The salinity, current density, product concentration, and pH conditions of the chloralkali process are vastly different from the conditions at a protected surface in the natural marine environment with current density on the order of 20μA/cm^2. Determination of reaction products and reaction selectivity under the applicable conditions is necessary to allow design of electrochemical antifouling systems. The present work utilizes a well characterized millimeter scale laminar flow electrochemical reactor and amperometric methods to determine reaction products, which yields reaction selectivity.

In order to evaluate the longevity of conductive antifouling coatings in marine applications accelerated aging methods are required. Standard procedures for rapid aging of marine paints involve exposure to ultraviolet light, moisture, temperature swings, and salt fog[5]. The two primary concerns for MHK applications are adhesion and electrochemical performance. The existing test regimens do not address electrochemical performance; therefore in addition to the standard tests accelerated aging methods using high over-potentials and current density have been developed. The electrode selectivity is characterized throughout the aging process, which enables prediction of performance losses.

Aged coatings have also been deployed in the marine environment to validate performance after aging. A standard procedure for washing test coupons to remove loosely adhered debris using consistent shear stress followed by visual and microscopic inspection has been developed as a means to standardize determination of antifouling efficacy.

[1] T. Matsunaga, T. Nakayama, H. Wake, M. Takahashi, M. Okochi, and N. Nakamura, "Prevention of marine biofouling using a conductive paint electrode," *Biotechnol. Bioeng.,* vol. 59, pp. 374-378, // 1998.

[2] T. Nakayama, H. Wake, K. Ozawa, N. Nakamura, and T. Matsunaga, "Electrochemical prevention of marine biofouling on a novel titanium-nitride-coated plate formed by radio-frequency arc spraying," *Appl. Microbiol. Biotechnol.,* vol. 50, pp. 502-508, // 1998.

[3] R. E. Perez-Roa, M. A. Anderson, D. Rittschof, C. G. Hunt, and D. R. Noguera, "Involvement of reactive oxygen species in the electrochemical inhibition of barnacle (Amphibalanus amphitrite) settlement," *Biofouling,* vol. 25, pp. 563-71, // 2009.

[4] J.-R. Huang, W.-T. Lin, R. Huang, C.-Y. Lin, and J.-K. Wu, "Marine biofouling inhibition by polyurethane conductive coatings used for fishing net," *J. Coat. Technol. Res.,* vol. 7, pp. 111-117, // 2010.

[5] *Paint and Coating Testing Manual : 15th edition of the Gardner-Sward handbook*: ASTM International, 2012.

Tangible Benefits Derived from the Application of the Hazard and Operability (HAZOP) Methodology for Process Safety Risk Assessments of Mobile Offshore and Land Drilling Units

David A. Jones
Atkins
920 Memorial City Way, Suite 700
Houston, Texas 77024
david.a.jones@atkinsglobal.com

Prepared for Presentation at
American Institute of Chemical Engineers
2014 Spring Meeting
10th Global Congress on Process Safety
New Orleans, LA
March 30 – April 2, 2014

UNPUBLISHED

Tangible Benefits Derived from the Application of the Hazard and Operability (HAZOP) Methodology for Process Safety Risk Assessments of Mobile Offshore and Land Drilling Units

David A. Jones
Atkins
920 Memorial City Way, Suite 700
Houston, Texas 77024
david.a.jones@atkinsglobal.com

Keywords: process safety, risk assessment, HAZOP, human factors, drilling operations, operational risk management

Abstract

This paper will demonstrate the tangible benefits derived from the practical application of a rigorous and systematic risk assessment methodology, the Hazard and Operability (HAZOP) study, that has been proven successful for nearly 50 years in the process industries, including upstream drilling and production design and operations.

The HAZOP process brings together a multi-disciplinary and interdependent team of subject matters experts from operations, engineering and process safety, representing the operating company, the drilling contractor, and equipment/service suppliers, to enable a comprehensive review of the technological and management systems for a drilling unit's design, operation and maintenance.

1. Introduction

Operational risk differs from other types of enterprise-wide risk because it is not used to generate profit. It involves the risk of direct or indirect loss resulting from inadequate or failed internal processes, people and systems, or from external events. Three of the seven official Basel II defined event types for operational risk (namely: workplace safety, damage to physical assets, and business disruption and systems failures) include health, safety, security, environmental (HSSE) and business continuity risks.[12] As a significant source of such risks in the process industries, process safety risks are derived from process hazards associated with the processing, storage, transport, and handling of hazardous substances.

The Hazard and Operability (HAZOP) methodology has been successfully applied as a process safety risk assessment tool for process systems in various industries, including oil and gas drilling, production, processing, and transportation.[3,14,16-18,20-22,25-27,29] This paper will focus on the application of the HAZOP methodology to oil and gas drilling operations (e.g. drilling, tripping, well control, well kill, cementing, etc.) for offshore and land-based units utilizing, for example, rotary, coiled tubing, and managed pressure drilling systems.

Specifically, examples will be presented for evaluating risk reduction opportunities applicable to surface gas and drilling fluid (mud) systems. These examples will highlight the tangible benefits derived from the application of the HAZOP methodology based on the author's global experience facilitating numerous HAZOP studies over the past 20 years for a variety of process systems in different industries, including mobile offshore drilling units (MODUs) (drillships, semisubmersibles, tender assisted barges, jack-ups) and land units (including Arctic rigs) in North America, Gulf of Mexico, Caribbean Sea, North Sea, and Middle East/North Africa, etc.

2. Process Safety – The Need for Assurance

Over the past few decades, major process safety incidents (e.g. Bhopal, Piper Alpha, Valdez, Macondo, etc.) have stressed the need for better management of process hazards and their associated risks [30]. As a result, various regulations[11,36] and industry standards and guidelines [3-4,14-16,18,20-23,25] have been developed to address the need for preventing such incidents and for providing to stakeholders (e.g. regulators, company management, insurers, the public, and the workforce) assurance that process safety risks are being properly managed to avoid or minimize loss to people, property, and the environment.

2.1 Process Hazards and Associated Risks – The Problem

"All risks to which the practice of safety applies derive from hazards: there are no exceptions" [31]. Frameworks from international standards and guidelines developed over the past few decades define a broadly applicable process for managing safety and other types of risks,[20-21,24-25,33] as follows:

- Risk Assessment
    - Risk Identification (e.g. hazard identification for safety applications)
    - Risk Analysis
        - Estimation of Severity of Consequences
        - Estimation of Likelihood of Event with Consequences
    - Risk Evaluation
- Risk Treatment
    - Risk Avoidance
    - Risk Transfer
    - Risk Control (i.e. provide or modify the controls)

This framework is consistent with a general definition for the safety profession: "The practice of safety is the identification, evaluation, and control of hazards to prevent or mitigate harm or damage to people, property, or the environment".[31] Process safety management (PSM) is defined as "the application of management systems to the identification, understanding, and control of process hazards to prevent process-related incidents and injuries. It is focused on the prevention of, preparedness for, mitigation of, response to, and restoration from catastrophic releases of chemicals or energy from a process associated with a facility."[13] At the heart of effective process safety management is a rigorous risk assessment and treatment process for identifying process hazards, analyzing and evaluating their associated risks, and controlling such risks to an acceptable (or tolerable) level (see Section 2.2).

Example process risk assessment methodologies from regulations, industry standards and guidelines include:[3, 14, 16, 18, 20-23, 25, 36]

- Bow Tie Analysis
- Event Tree Analysis (ETA)
- Failure Modes and Effects Analysis (FMEA)
- Fault Tree Analysis (FTA)
- **Hazard and Operability (HAZOP) Study**
- Hazard Identification (HAZID) Study
- Layers of Protection Analysis (LOPA)
- Quantitative Risk Assessment (QRA)
- Safety Integrity Level (SIL) Analysis
- What-If / Checklist Study

For drilling units, risk assessments may be required to address: (1) design decisions for specific equipment (e.g. for blowout preventers (BOPs));[1] (2) assurance processes involving the manufacture and delivery of quality products and services (e.g. API Spec Q1 & Q2);[7-8] or (3) the process design for offshore drilling and production facilities and operations (API RP 14J).[3]

2.2 Risk Control – The Solution

Effective PSM systems are designed with key management processes for controlling risks at three opportune times:[28]

- *Before* an incident – prevention through design and operational discipline in order to identify, avoid, eliminate or minimize risks to prevent occurrence;
- *During* an incident – emergency response in order to prepare for and respond to an incident; and business continuity in order to maintain or restore operations while minimizing losses; and
- *After* an incident – incident investigation, root cause analysis, and implementation of lessons learned in order to prevent recurrence.

Predictive process risk assessments are most appropriately applied to prevent incidents, by designing risk controls into the technological and management systems, ideally at the engineering design stage, or later as recommended risk reduction measures.

If we consider the axiom that 'if you are not in control, then you are out of control'; then a challenging question would be whether or not the specified risk controls for a particular process design are sufficient to be in control (i.e. to prevent or otherwise mitigate an event). And since we cannot control what we don't understand, the only way to answer the question with confidence for process safety is to rigorously identify the process hazards, then analyze and evaluate their risks.

2.3 *Hazard and Operability (HAZOP) Study – A Proven Methodology*

The HAZOP study is one of the most systematic and rigorous methodologies for identifying and analyzing process hazards and operability issues, and has proven successful for nearly 50 years in the process industries, including upstream drilling and production design and operations. [3,14,16-18,22,25,29] The prominence of the HAZOP methodology is evidenced by its reference in various major offshore investigation reports over the past few decades (e.g. Piper Alpha, Deepwater Horizon).[10,19]

A HAZOP study brings together the collective knowledge and experience base of a small team in a short amount of time to deepen the understanding of process hazard scenarios of a particular process design under review (see Section 2.3.1). The HAZOP methodology is employed for studying the design intent of the process, identifying process hazards (causes and consequences), evaluating the effectiveness of existing safeguards, estimating the levels of risks, and justifying the need for reducing the risks by recommending actions for improvement. Thus, HAZOP studies are well suited for conducting process risk assessments and generating recommendations for risk control.

The key benefits of utilizing the HAZOP methodology are that: (1) its selection and allocation of the resources necessary to support the study reflects management's commitment for a proactive and thorough approach to risk management through continual improvement; and (2) its prediction nature offers an opportunity for early learning and prevention without having to experience the hazard scenario in the real world (i.e. rather than depending solely on incorporating lessons learned from previous incidents).

2.3.1 Hazard Scenarios – Postulated Hazardous Events

The HAZOP methodology aims to identify and analyze process hazards associated with events defined by causes and the resulting adverse consequences. These cause-consequence event sequences, along with associated safeguards for prevention (of the causes) and mitigation (of the consequences), and risk ranking of the severity (of the consequences) and likelihood (of the events), are documented in HAZOP worksheets as hazard scenarios. Figure 1 shows a simple example of a hazard scenario documented in a HAZOP worksheet.

HAZOP *causes* help to answer "what can go wrong (and how often can it happen)?", addressing human errors, equipment malfunctions, and external events (see Section 2.3.2). HAZOP *consequences* help to answer "what are the potential effects (and how severe can they be)?", addressing impacts to people, property, the environment, assets, continuity and reputation (see Section 2.3.3). HAZOP *safeguards* help to answer "what existing risk controls exist (and how effective are they)?", addressing barriers for prevention, detection, control, mitigation, and response (see Section 2.3.4). HAZOP *risk ranking* helps to answer "what is the risk level (and is further risk reduction warranted)?", addressing the severity of the worst credible consequences and the likelihood of the hazardous event, accounting for a reasonable evaluation of the effectiveness of the existing safeguards (see Section 2.3.7).

Figure 1: Example of Hazard Scenario in HAZOP Worksheet

Node: 3, Mud Gas Separator	**Parameter:** Level	**Design Intent:** To separator gas & mud during well control

Deviation	**Cause**	**Consequences**	**Safeguards**	**S**	**L**	**R**	**Recommendations**
Higher Level	Manual choke set too high in error	Potential to overfill the Mud Gas Separator resulting in a release of mud onto the deck or overboard, with possible harm to personnel in the vicinity and impact to the environment	Procedure P-002 for verifying valve alignments and choke settings for the Choke and Kill Manifold (CKM-2)	3	C	M	Consider providing a means to detect high level in the Mud Gas Separator (e.g. level or pressure indicator). Consider assigning a "watch" to observe the MGS atmospheric vent during Well Control operations.

Throughout the development of the hazard scenarios, the HAZOP team is also challenged to incorporate organizational and human factors (see Section 2.3.5) and process integrity (see Section 2.3.6) into the process hazard analysis, in order to address the human element and hardware aspects, respectively.

The key benefits of analyzing process hazards using hazard scenarios are that the HAZOP methodology can help to assure: (1) completeness of defining the breadth of possible problems through systematic hazard identification; and (2) comprehensiveness of evaluating in depth possible solutions through a rigorous evaluation of the current design and treatment of its potential risks.

Example *process hazard scenarios* for drilling units include the potential: to release drilling fluid from the bell nipple due to an induced kick (loss of hydrostatic) or a blowout; to overfill a mud gas separator (MGS) due to a choke washout; to blowby gas to the shakers due to a loss of the liquid seal (or leg) in the MGS; and to rupture a hose and release high pressure mud due to deadheading a mud pump.

2.3.2 Causes – Derived from Process Deviations

The HAZOP methodology utilizes an *inductive* approach which starts with a possible cause and asks what consequences (or effects) would happen it were to occur. This approach is different than an *deductive* approach such as that utilized by the Fault Tree Analysis methodology which starts with a consequential event (or top event) and asks what would be the causes for it to occur.[14, 23]

Unlike using a free-style brainstorming technique such as What-If Analysis, the HAZOP methodology uses process deviations to stimulate the team's efforts to postulate possible causes. Process deviations are defined as conditions of operation outside the design envelope or deviations from procedures (defined by relevant process safety information); and are derived from the application of selected guidewords to the relevant process parameters in the design. For example, HAZOP guidewords such as No, Less, More, Reverse, etc. combined with a process parameter such as Flow, Pressure, Temperature, Level, etc. may yield process deviations such as No Flow, Lower Pressure, Higher Level, etc. Once the design intents are defined for the section of the process (called HAZOP node) under review, then the HAZOP facilitator would challenge the team to identify possible causes for a specific process deviation. More detailed step-by-step procedures for a HAZOP study are defined in industry standards and guidelines.[14,18,22-23,29]

Three broad categories of causes typically treated within the hazard scenarios include: human errors, equipment malfunctions, and external events. The HAZOP team's knowledge of the process design and its operation is crucial for defining the causes for the hazards scenarios. This knowledge base can be improved through proper team composition, as well as pre-HAZOP training. Also, the causes of previous incidents are considered a valuable source of information during the HAZOP and help to validate the team's analysis.

The key benefits of striving to identify all possible causes of process deviations are that: (1) it stimulates "thinking outside the box" from what goes right to what could go wrong (i.e. reverse engineering); and (2) it captures both commonly encountered scenarios (including less frequent but known examples from previous incidents) as well as newly postulated scenarios that may never have occurred.

Example *causes* for drilling units include: manual valve in the choke and kill manifold is open (or closed) in error; hydraulically operated valve on the kill line malfunctions closed (or open); pressure relief valve (PRV) on the high pressure mud pump is set in error or leaks through; kick or influx from the well.

2.3.3 Consequences – Extended to Worst Credible Conditions

Consequences for each hazard scenario are developed from the cause as the initiating event. For most scenarios, the physics (and chemistry) of the process fluids are discussed to analyze and understand what happens next. Examples include the physical effects of gravity, pressure changes, momentum changes, dispersion (or accumulation), blast effects, fire radiation effects, etc. Drawing from their engineering and operational knowledge and expertise, the HAZOP team describes the sequential changes and subsequent events which are the effects of the previous condition. For example, consequences may be documented in terms of the physical mechanism (e.g. deadheading of a pump) and resulting equipment effect (e.g. seal damage), resulting in the physical phenomenon of a loss of containment with a destination (e.g. high pressure spray upwards; mud spill onto the deck), and a determination of the potential impacts to people, property and the environment.

A common practice in applying the HAZOP methodology is to carry out the hazard scenario to its ultimate (or worst credible) consequences; this is, by assuming no credit for the existing safeguards. The term "credible" means what is physically possible, not whether such consequences have ever occurred or might occur. In this manner, we can look beyond each layer of protection provided by the safeguards in order to help define and understand their purpose. The consequences of previous incidents are considered a valuable source of information during the HAZOP which aids in the brainstorming exercise and adds credibility to the analysis of the hazard scenarios.

The key benefits of analyzing the worst credible consequences are that: (1) it raises awareness and documents how hazardous events can occur; and (2) it answers what could happen if no safeguards were in place. Settling for anything short of the ultimate consequences underestimates the potential process hazards and their risks.

Example *consequences* of process safety events for drilling units with possible impact to people, property or the environment, include the potential: to release drilling fluid from the bell nipple or overwhelm the mud flowline due to an induced kick (loss of hydrostatic) or a blowout resulting in an environmental spill; to overpressure a mud gas separator (MGS) due to a choke being set in error, resulting in a release of drilling fluid and gas; for a loss of liquid level in the MGS, resulting in a flammable gas release through the anti-siphon vent with possible fire/explosion; to collapse the well resulting in stuck pipe; to pressure up the wellbore resulting in fracturing the well with a possible underground blowout; and to rupture a hose and release high pressure mud towards persons in the vicinity.

2.3.4 Safeguards – Verified and Evaluated

Safeguards are also referred to as risk controls, risk reduction measures, barriers, lines of defense, and layers of protection. They include both engineering and administrative controls addressing the full range of hierarchy of controls, including: inherently safer design (ISD) principles, prevention, detection, control, mitigation, and emergency response.[15,20,21,28,33]

The HAZOP study identifies and documents existing safeguards for each hazard scenario. Many equipment and instrumentation safeguards are shown on the process and instrumentation diagrams (P&IDs) under review in the HAZOP study, but many others must be verified through review of relevant design or operating documents (e.g. equipment data sheets, operating procedures) compiled as process safety and environmental information and supplied to the HAZOP team for the study.

Safeguards are typically defined by standards, codes, and regulations and include components or elements of the both the technological system (i.e. the engineered process design) and the operational or health, safety & environmental (HSE) risk management system. By understanding the function of each relevant safeguard, the HAZOP team can then evaluate their availability, adequacy, and effectiveness for the respective hazard scenarios.

The key benefits of verifying and evaluating the existing safeguards are that: (1) it enhances the development of hazard scenarios by justifying safeguards in terms of their role in the prevention of causes and the mitigation of consequences (for example, probing "why a check valve exists in the line" or "what worse could happen if we didn't measure the flow"); and (2) it enables a more complete and meaningful evaluation of their overall effectiveness for a specific hazard scenario, and their contribution to the current risk level.

Example *safeguards* for drilling units include:

- Inherently safer design features – not allowing (by design) direct flow from the well (annulus) to the MGS as a choice for emergency diverting; remote location of a secondary containment tank for diverting; design for 100% secondary containment.

- Prevention – competence assurance program (e.g. well control training certification) to minimize human errors; preventive maintenance program (e.g. periodic pressure and leak testing of high pressure components) to minimize equipment malfunctions; hazardous area classification to prevent ignition sources.
- Detection – indication and alarms for process parameters such as flow (e.g. mud trough or flowline), pressure (e.g. standpipe manifold), level (e.g. active mud tanks), temperature (e.g. well fluids); mud weight (e.g. mud logging); gas concentrations in air (e.g. gas detection).
- Control – manual or automatic shutdown of mud pumps (e.g. pressure interlocks); diverting flow overboard through a diverter; use of blowout preventer (BOP) to isolate well fluids.
- Mitigation – secondary containment (e.g. deck drain systems; cuttings bins); fire or blast walls.
- Response – emergency shutdown system; emergency response plans and drills.

2.3.5 Organizational and Human Factors – Addressed in Each Hazard Scenario

The HAZOP methodology places each team member at the center of attention in a high-impact learning environment. Adults learn best by doing, and the HAZOP methodology allows the team to learn by postulating "what-if scenarios". A well facilitated HAZOP study integrates organizational and human factors into the analysis by engaging the team as humans who find themselves in the operational loop with a vested interest in performing their job, protecting themselves in the field, and controlling their decisions and actions to help prevent or mitigate or respond to an event.

General examples for addressing *organizational and human factors* within the HAZOP worksheets for each hazard scenario include: (for causes) human errors in operation and maintenance; (for consequences) potential exposure and impact to personnel health and safety; and (for safeguards) human-machine interface design features; competence assurance including selection, orientation, training, development of personnel and contractors; programs and procedures with human decisions and actions for preventive maintenance, safe work practices and permit systems, process monitoring, response to alarms and intervention for controlling process upsets, and emergency response.[33,34,35]

2.3.6 Process Integrity – Addressed in Each Hazard Scenario

A well-organized HAZOP study integrates aspects of asset integrity into the analysis, particularly process integrity (e.g. containment, pressure control, corrosion control, flow assurance, etc.), by addressing piping, mechanical, instrumentation and controls, and other disciplines. During the study, process integrity factors should be anticipated throughout the asset life cycle: design, installation, maintenance, and operation of the facilities, equipment and tools. Industry standards are an excellent source of information for addressing process integrity and associated design requirements of the process and its safeguards.[1-3,5-6,9,33]

General examples for addressing *process integrity* within the HAZOP worksheets for each hazard scenario include: (for causes) equipment malfunctions; (for consequences) potential impact to equipment or facilities; and (for safeguards) programs and procedures with conditions for and response by safety-critical equipment and systems for prevention, detection, control, mitigation, and emergency response.

2.3.7 Risk Levels – Estimated and Evaluated for Acceptability

Determination of the risk level for each hazard scenario can be achieved by estimating the worst credible impact to people, property and the environment (i.e. based on consequences without the benefits of any safeguards) and by estimating the likelihood of the scenarios (i.e. based on the probability or frequency of the initiating cause and by accounting for a reasonable level of effectiveness of the existing safeguards).

A HAZOP study often employs a risk assessment matrix to assist the team in semi-quantitatively ranking the estimated risks for each hazard scenario. The results of the estimated risk are typically used as a guide to determine whether recommendations are desired. Many organizations using a risk assessment matrix also have screening criteria to evaluate the acceptability (or tolerability) of the risk and to evaluate the need for further risk reduction.[20-21, 25] An example risk assessment matrix, with criteria for recommendations, is shown in Figure 2.

Figure 2: Example Risk Assessment Matrix

Severity Rating	Consequences				Likelihood Rating				
	People	**Environment**	**Assets**	**Reputation**	**A** Has never occurred in industry	**B** Has occurred in industry	**C** Has occurred in Company	**D** Has occurred frequently in Company	**E** Has occurred frequently at Site
5	Multiple Fatalities	Massive Effect	Extensive Damage	International Impact					
4	Single Fatality	Major Effect	Major Damage	National Impact					
3	Major Harm	Local Effect	Local Damage	Considerable Impact					
2	Minor Harm	Minor Effect	Minor Damage	Limited Impact					
1	Slight Harm	Slight Effect	Slight Damage	Slight Impact					
0	No Harm	No Effect	No Damage	No Impact					

Key	Risk Level	Guidance for HAZOP Recommendations
	High	Incorporate additional or improved risk reduction measures
	Medium	Consider additional or improved risk reduction measures
	Low	Manage existing safeguards for continual improvement

The key benefits of estimating the risk for each hazard scenario are that: (1) it provides a consistent basis for determining the need for improving or adding safeguards to further reduce the risk; and (2) it provides a means to compare the relative risks between each scenario as well as the relative significance of certain causes and consequences or the criticality of certain safeguards.

2.3.8 Recommendations – Proposed for Action

HAZOP study results typically include recommendations for consideration by management to treat the risks (i.e. by reducing the severity of consequences and/or the likelihood of occurrence), to improve operability, to conduct further analyses beyond the scope of the study, or to verify design or operating data from process safety and environmental information not available during the study. Follow-up on proposed actions from recommendations, including their resolution, implementation (e.g. designing a solution), and tracking their closure, are addressed following the HAZOP study. Such follow-up and the decisions for actions are often carried out by others not involved on the HAZOP team.

The key benefits of making recommendations are that: (1) it provides a means and technical justification for changes in the process design or its operation in order to reduce risks and improve operability; and (2) it provides a means to document considerations for and comparison to other hazard scenarios as well as process designs (whether these are between future design alternatives or between different existing drilling units).

Example *recommendations* for drilling units include: verifying the sizing of process equipment (e.g. MGS, manifold, and piping/hose pressure ratings) and devices (e.g. pressure relief valves on mud pumps); developing or revising operating procedures to address proper alignment of safety-critical valves (e.g. manifold valves); developing or revising preventive maintenance procedures (e.g. pressure testing); providing alarms for critical pressure, flow and level indicators, including redundancy where justified (e.g. mud flowline, MGS, mud tanks, mud tanks); installing closed-circuit televisions (CCTVs) in remote areas of the drilling rig where potential spills can be monitored (e.g. cuttings collection area); incorporating into well control training and operational drills events and actions required for managing high-risk hazard scenarios (e.g. circulating out gas from a kick).

3. Benefits of a HAZOP Study

Various tangible benefits can derived from the practical application of the HAZOP methodology as a tool for process risk assessment. The HAZOP study adds value throughout its process (i.e. the journey) and from its results (i.e. the destination).

3.1 Benefits from the HAZOP Study Process

The HAZOP process brings together a multi-disciplinary and interdependent team of subject matters experts from operations, engineering and process safety, to enable a comprehensive review of the technological and management systems for a facility's design, operation and maintenance. Below are five example benefits from this proven and popular team-based activity.

3.1.1 Increased Workforce Participation

The HAZOP process enables team members to actively participate in challenging both the technological and management systems of the process design and operation under review without placing blame on individuals. Facilitating such participation in a learning environment adds value via the team's shared knowledge and understanding of what can go wrong and empowers them to make recommendations for reducing risks.[26]

Experience has shown that setting an expectation that all team members can contribute to the sessions from a position of equal status often results in very productive environment.

3.1.2 Improved Interface Management

The HAZOP process engages multiple parties responsible for the process design and its operation and maintenance. For drilling units, this team-based approach typically includes representatives from the operating company, the drilling contractor, and equipment/service suppliers, and also engineering contractors and vendors for capital projects. A team-based HAZOP session encourages and facilitates constructive communications and coordination across organizational interfaces.

From experience, in a majority of studies, especially those for capital projects involving a new or modified process design, the HAZOP team comprises members that have come together on the project for the first time, thereby providing an opportunity for teambuilding that extends beyond the study into construction, commissioning, and start-up activities.

3.1.3 Drilling Unit-Specific Awareness and Understanding

Because we don't experience process safety-related incidents frequently, we might not "see" process hazards nor perceive their potential risks, not to mention how they could occur and how to prepare to respond to them. The HAZOP process, through the systematic use of predictive scenario-based hazard identification and risk assessment, raises awareness of team members about the process design intents and operational philosophies, and helps them achieve an understanding of what can go wrong and why, and how to prevent or mitigate such events. As such, participation in a HAZOP study can be considered an important complement to on-the-job training (OJT) for the drilling unit.

A HAZOP study can provide the team with a key takeaway; i.e. hopefully a different way of looking at their drilling unit. Experience in facilitating numerous HAZOP studies has over a twenty-year period generated nearly 100% of time the following typical response from one or more team members: "I've learned more about my system/process/operation during a weeklong session than I have working on it for years".[26]

3.1.4 Shared Ownership for Operational Risk Management

The HAZOP process, through a team-based approach, promotes a shared responsibility for and ownership of the decisions that affect the design, operation, and maintenance of the process/system under review. Such ownership improves the effectiveness of managing process safety as part of the drilling unit's operational risk management strategy.

From experience, the adage "once you *see* it, you *own* it" applies to the identified process hazards and the recommendations of the HAZOP team gained by consensus during the study, for both the current and future design and its operation.

3.1.5 Strengthened Culture of Prevention

The overall HAZOP process, including planning and preparation, team participation, documentation, and follow-up actions for recommendations, coupled with a robust management of change (MOC) process, strengthens a culture of prevention through continual improvement of the process design and its operation. Leadership by using a proactive, predictive approach and management commitment to allocating sufficient resources for conducting a HAZOP study, demonstrate to stakeholders the importance of process risk management.[32]

From experience, organizations which go beyond minimum compliance with regulatory requirements and strive for prevention, can sustain a culture of continual improvement in operational risk management performance.

3.2 Benefits from the HAZOP Study Results

The HAZOP process delivers results through both the study documentation itself and how the contents of the documentation are used. Below are five example benefits derived from the results of this rigorous and systematic process.

3.2.1 Purposeful Risk Reduction

The HAZOP process documents the results of the primary purpose of the study which is to recommend further risk reduction measures based on an assessment of the identified hazards and evaluation of their associated risks based on existing safeguards. Such recommendations for reducing the severity of consequences and/or the likelihood of occurrence typically address missing or inadequate safeguards as layers of protection, based on the hierarchy of controls and specific to each hazard scenario.

Examples of *risk reduction measures* include recommendations for provision of missing or inadequate safeguards (see Sections 2.3.4 and 2.3.8).

3.2.2 Supplemental Design Verification and Validation

The HAZOP process serves to document the results of verification (i.e. does the design meet requirements) and validation (i.e. does the design fulfill its intended purpose both from a safety and operability standpoint). As such, the HAZOP results provide a quality check as a supplemental safety design review.

Examples of *verification* for drilling units include: confirmation of the existing, new or modified design basis (including the quality of process safety and environmental information) such as for pressure relief valve settings and hose pressure ratings; and confirmation of required management system documents such as operating procedures and training covering all operating modes.

Examples of *validation* for drilling units include: confirmation of the design's intended use during operation and maintenance (i.e. are the safeguards suitable to control risks for all credible hazard scenarios) such as determining that the MGS is capable of handling certain pressure conditions for an identified hazardous event; and confirmation that the process design and its equipment is suitable for certain reverse flow scenarios which are credible but not intended; and confirmation that emergency response plans and drills are valid for all hazard scenarios.

3.2.3 Assurance for Compliance and Conformance

The HAZOP study documentation provides a basis for demonstrating that certain requirements of process safety regulations and/or company expectations have been met; i.e., providing assurance for compliance and conformance, respectively. HAZOP worksheets also serve as records for future compliance audits, as well as a repository for hazard scenarios which can be referenced for future process safety-related incidents.

An example of *compliance assurance* for drilling units includes: reports for gas and mud systems hazard analyses and their periodic revalidation as required by regulations for Safety and Environmental Management Systems (SEMS).[11]

An example of *conformance assurance* for drilling units includes: reports for the analyses of new hazards before installing a managed pressure drilling (MPD) system on an existing drilling rig in order to address the changes to the existing systems as required by management of change (MOC) procedures and pre-startup safety reviews.

3.2.4 Enhanced Management Systems

HAZOP study reports with detailed worksheets provide an excellent basis for developing documents for other PSM/SEMS elements, including standard operating procedures, training programs, emergency response plans, and equipment integrity plans and procedures. Often, subject matter experts may attend HAZOP sessions as observers who take notes to evolve these work products.

Examples of *management system enhancements* for drilling units include: developing manual valve alignment tables for various operating modes of the choke and kill manifold and related mud handling systems; identifying high risk scenarios to enhance existing plans for well control and emergency drills; and preparation of safety-critical equipment lists for preventive maintenance programs.

3.2.5 Continual Improvements for Operability

The HAZOP study methodology was originally devised to address improvements for operability of process systems.[29] Although most HAZOP studies stress the importance of identifying and analyzing process hazards, a well defined scope can yield additional benefits of improved operability. Such opportunities for improving the performance of the process/system may include reduction of non-productive time (NPT) related to equipment failures; reduction of reprocessing and associated waste of energy, time or product; or reduction of operating costs.

Examples of *operability improvements* for drilling units include: decreased cycle time for tripping; improved mud quality and reduction of chemical usage; prevention of significant NPT and associated financial losses due to downhole equipment damage (e.g. stuck pipe), formation damage, and mud losses related to process safety-related incidents.

4. Conclusions

This paper has presented several tangible benefits resulting from the practical application of the HAZOP methodology that are based on its demonstrated success in the process industries for nearly 50 years. This paper has also provided several examples specific to the review of the technological and management systems for a drilling unit's design, operation and maintenance. These benefits and examples have been derived from both the implementation of the HAZOP process and the HAZOP study results produced.

A HAZOP study is much more than a simple "check-the-box" activity to satisfy project management objectives or regulatory requirements; it is a rigorous, systematic, predictive and proven tool for process safety risk assessments undertaken as a core process within a proactive operational risk management framework.

Successful application of the HAZOP methodology adds values to the bottom line, helps to protect the environment, and most importantly, is instrumental in preventing the loss of lives!

5. References

[1] API. *Blowout Prevention Equipment Systems for Drilling Wells*, API Std 53, Fourth Edition, American Petroleum Institute, Washington DC, November 2012.

[2] API. *Guide for Pressure-Relieving and Depressuring Systems*, API Std 521, Fifth Edition, American Petroleum Institute, Washington DC, January 2007.

[3] API. *Recommended Practice for Design and Hazard Analysis for Offshore Production Facilities*, API RP 14J, Second Edition, American Petroleum Institute, Washington DC, April 2001 (Reaffirmed March 2007).

[4] API. *Recommended Practice for Development of a Safety and Environmental Management Program for Offshore Operations and Facilities*, API RP 75, Third Edition, American Petroleum Institute, Washington DC, May 2004 (Reaffirmed May 2008).

[5] API. *Recommended Practice for Well Control Operations*, API RP 59, Second Edition, American Petroleum Institute, Washington DC, May 2006 (Reaffirmed January 2012).

[6] API. *Specification for Choke and Kill Systems*, API Spec 16C, First Edition, American Petroleum Institute, Washington DC, January 1993 (Reaffirmed July 2001).

[7] API. *Specification for Quality Management System Requirements for Manufacturing Organizations for the Petroleum and Natural Gas Industry*, API Spec Q1, Ninth Edition, American Petroleum Institute, Washington DC, June 2013.

[8] API. *Specification for Quality Management System Requirements for Service Supply Organizations for the Petroleum and Natural Gas Industries*, API Spec Q2, First Edition, American Petroleum Institute, Washington DC, December 2011.

[9] API. *Recommended Practice for Diverter Systems Equipment and Operations*, API RP 64, Second Edition, American Petroleum Institute, Washington DC, October 2001 (Reaffirmed January 2012).

[10] BP. *Deepwater Horizon Accident Investigation Report*, BP, September 8, 2010. http://www.bp.com/liveassets/bp_internet/globalbp/globalbp_uk_english/incident_response/STAGING/local_assets/downloads_pdfs/Deepwater_Horizon_Accident_Investigation_Report.pdf (accessed Dec. 26, 2013).

[11] BSEE. Workplace Safety Rule, 30 CFR Part 250, "Oil and Gas and Sulphur Operations in the Outer Continental Shelf, Subpart S – Safety and Environmental Management Systems (SEMS)", §250.1911 (reference to Hazards Analysis), Bureau of Safety and Environmental Enforcement (BSSE), *Federal Register*, October 15, 2010, (75 FR 199).

[12] Basel Committee for Banking Supervision. *Principles for the Sound Management of Operational Risk and the Role of Supervision*, ISBN: 92-9131-857-4, Bank for International Settlements, Basel, Switzerland, June 2011.

[13] CCPS. *Guidelines for Auditing Process Safety Management Systems*, 2nd Edition, ISBN: 978-0-470-28235-9, Center for Chemical Process Safety (CCPS) of the American Institute of Chemical Engineers (AIChE), New York, New York, 2011.

[14] CCPS. *Guidelines for Hazard Evaluation Procedures*, 3rd Edition, ISBN: 978-0-471-97815-2, Center for Chemical Process Safety (CCPS) of the American Institute of Chemical Engineers (AIChE), New York, New York, 2008.

[15] CCPS. *Inherently Safer Chemical Processes: A Life Cycle Approach*, 2nd Edition, ISBN: 978-0-471-77892-9, Center for Chemical Process Safety (CCPS) of the American Institute of Chemical Engineers (AIChE), New York, New York, 2008.

[16] CIA. *A Guide to Hazard and Operability Studies*, Chemical Industries Association (CIA), London, UK, (1977) 1992.

[17] Comer, P.J, J.S. Fitt, R. Østebø. "A Driller's HAZOP Method," Society of Petroleum Engineers (SPE) European Petroleum Conference, London, UK, October 20-22, 1986.

[18] Crawley, Frank et al. *HAZOP: Guide to best practice*, 2nd edition, ISBN: 978 0 85295 525 3, Institute of Chemical Engineers (IChemE), Rugby, UK, 2008.

[19] Cullen, The Hon. Lord William Douglas. *The Public Inquiry into the Piper Alpha Disaster, Presented to Parliament by the Secretary of State for Energy by Command of Her Majesty*, ISBN 0101113102, H.M. Stationery Office, London, UK, November 1990.

[20] IADC. *Health, Safety and Environmental Case Guidelines for Land Drilling Units*, Issue 1.01, International Association of Drilling Contractors, Houston, Texas, July 2009.

[21] IADC. *Health, Safety and Environmental Case Guidelines for Mobile Offshore Drilling Units*, Issue 3.4, International Association of Drilling Contractors, Houston, Texas, November 2011.

[22] IEC. *Hazard and operability studies (HAZOP studies) – Application guide*, International Standard IEC 61882:2001, First edition, International Electrotechnical Commission (IEC), Geneva, Switzerland, May 2001.

[23] IEC. *Risk management – Risk assessment techniques*, International Standard IEC/ISO 31010:2009, Edition 1.0, International Electrotechnical Commission (IEC), Geneva, Switzerland, November 2009.

[24] ISO. *Risk management – Principles and guidelines*, International Standard ISO 31000:2009(E), First Edition, International Organization for Standardization (ISO), Geneva, Switzerland, November 15, 2009.

[25] ISO. *Petroleum and natural gas industries – Offshore production installations – Guidelines on tools and techniques for hazard identification and risk assessment*, International Standard ISO 17776:2000, First Edition, International Organization for Standardization (ISO), Geneva, Switzerland, 2000.

[26] Jones, David A. (presenter and panelist). "Deepening the Workforce's Understanding of Process Safety – Using Team-Based Hazard and Operability (HAZOP) Studies", Drilling Health, Safety, Environment (HSE) & Training Europe 2013 Conference & Exhibition, International Association of Drilling Contractors (IADC), Amsterdam, The Netherlands, September 25-26, 2013.

[27] Jones, David A. "Enhancing HAZOP Studies – Adding Value by Integrating Hazard Analyses into the Process Safety & Environmental Management System (PSM/SEMS)", Decomworld Offshore Process Safety Conference, Houston, Texas, September 12, 2013.

[28] Jones, David A. "Using a Risk-Based Process to Design Inherently Safer and More Reliable Technological Systems", 2006 International Symposium, Mary Kay O'Connor Process Safety Center, College Station, Texas, October 24-25, 2006.

[29] Kletz, Trevor A. *Hazop and Hazan,* 4th Revised edition (paperback), ISBN: 978-0852954218, The Institute of Chemical Engineers (IChemE), Rugby, UK, 2006.

[30] Mannan, Sam (Editor). *Lee's Loss Prevention in the Process Industries: Hazard Identification, Assessment and Control*, Third Edition, ISBN: 0-7506-7555-1, Elsevier Butterworth-Heinemann, Burlington, MA, 2005.

[31] Manuele, Fred A. *On the Practice of Safety*, Fourth Edition, ISBN: 978-1-118-47894-3, John Wiley & Sons, Inc., Hoboken, New Jersey, 2013.

[32] OGP. *A guide to selecting appropriate tools to improve HSE culture*, Report No. 435, The International Association of Oil and Gas Producers (OGP), London, UK, March 2010.

[33] OGP. *Deepwater Wells – Global Industry Response Group recommendations*, Report No. 463, The International Association of Oil and Gas Producers (OGP), London, UK, May 2011.

[34] OGP. *Human factors engineering projects*, Report No. 454, The International Association of Oil and Gas Producers (OGP), London, UK, August 2011.

[35] OGP. *Risk Assessment Data Directory – Human factors in QRA*, Report No. 434-5, The International Association of Oil and Gas Producers (OGP), London, UK, March 2010.

[36] OSHA. "Process safety management of highly hazardous chemicals", 29 CFR 1910.119, (e), Process Hazard Analysis, Occupational Safety and Health Administration (OSHA), *Federal Register*, February 24, 1992 (57 FR 6356).

The Regulation of Offshore Process Safety

Jerad M. Denton
U.S. Chemical Safety and Hazard Investigation Board
Washington, D.C. 20037
jerad.denton@csb.gov

Prepared for Presentation at
American Institute of Chemical Engineers
2014 Spring Meeting
10th Global Congress on Process Safety
New Orleans, LA
March 30 – April 2, 2014

The Regulation of Offshore Process Safety

Jerad M. Denton
U.S. Chemical Safety and Hazard Investigation Board
Washington, D.C. 20037
jerad.denton@csb.gov

Keywords: process safety, regulation, SEMS, safety case, performance-based, prescription, best practices, offshore

Abstract

In the aftermath of the Macondo incident, the U.S. regulatory framework concerning offshore process safety has been debated, reviewed, and revised in the hopes of reducing the likelihood of a similar occurrence. In an effort to contribute to this critically important discussion, the U.S. Chemical Safety Board (CSB) held a public hearing titled "Regulatory Approaches to Offshore Oil and Gas Safety." [1] The hearing was attended by international experts on safety practices and policies, members of the industry, regulators, unions, and stakeholders. Additionally, the CSB has extensively researched guidance and international regulation concerning offshore process safety through interviews, outreach efforts, and a host of other methods. The CSB will provide an overview of its research and explore various approaches to the regulation of offshore process safety.

1. Introduction

The Macondo incident and other similar accidents have had devastating consequences around the world. With the occurrence of such accidents comes the obligation to critically examine offshore drilling practices to make them safer and more reliable. To that end, the U.S. Chemical Safety and Hazard Investigation Board (CSB) is currently examining the Macondo incident in order to issue meaningful recommendations aimed at preventing future offshore accidents in U.S. waters. As part of this examination, the CSB has conducted extensive research on the U.S. regulation of offshore process safety.

The regulatory frameworks that the CSB has examined are located on a spectrum with prescription and performance-based regulation at either end. The prescriptive approach details specific legislative requirements in operations. The performance approach requires the regulated party focus on attaining a specified goal. The manner used in reaching the goal is generally left to the regulated party, but requires demonstration that the goal has been achieved. All regimes are a hybrid of these two approaches, yet each can be classified as more prescriptive or more performance-based.

The CSB looked to international regulatory frameworks to determine whether improvements could be made in the U.S. The United Kingdom, Norway, and Australia were examined. Each provided a view of a more performance-based approach to regulation. Such approaches provide industry, workforce, and the regulator with valuable insights into how each can strive to continuously improve upon the regulation of offshore safety.

The purpose of this discussion is to provide a glimpse of the CSB's work relating to the Macondo incident, particularly to the current U.S. regulation of offshore process safety. This paper should be viewed as a means to generate discussion concerning the regulation of offshore safety. The CSB will be publishing its Macondo investigation report, including its analysis of the U.S. offshore regulatory framework, this year.

2. Regulatory Approaches to Offshore Safety

Before looking at various regulatory regimes, it is helpful to have a general understanding of ways in which a legislature can approach regulation. There are two approaches which exist at opposite ends of the regulatory spectrum: prescriptive and performance-based regulation. An understanding of these approaches provides insights into how a regulatory framework is structured and enforced.

2.1 Prescriptive Regulation

A regulatory framework that is classified as prescriptive specifies technical or procedural requirements that must be met by industry. The focus of the prescriptive regulation is ensuring conformity with specified requirements. An example of such an approach would be a requirement that guardrails on scaffolding be 42 inches in height.

2.1.1 Benefits

Prescriptive regulation is best utilized in areas where best practices can be clearly defined, there is little need or potential for innovation, and where deviation from requirements could create unacceptable risks to the environment or human health. In other words, industries that benefit from this type of regulation are industries in which the risks are very well understood and there is little to no change in the industry. [2] Prescriptive regulation can also prove beneficial in industries where a standard floor of compliance is required of all parties. Such regulations have the added benefit of being both relatively simple to implement and small business friendly.

2.1.2 Challenges

Prescriptive regulations, however, suffer from their inability to prevent new types of accidents that may appear in the future. This is due to the inflexibility in prescriptive regulation. Given the legislative hurdles involved in promulgating new regulations in the U.S., the regulator can face difficulties in attempting to keep regulatory pace with changes in complex areas. Furthermore, given the rigid requirements of prescriptive regulation, innovation is also generally stymied. The requirement that the regulated parties comply with prescriptive regulation can also create a culture of compliance in which industry is more concerned with satisfying standards over consideration of the prioritization of an operation's risks. Prescriptive regulation can also generate disputes about whether certain regulations actually apply in a particular situation. These issues are especially problematic in industries with complex technologies and processes such as the chemical sector.

2.2 Performance-Based Regulation

Performance-based regulation involves specifying the goal that is to be attained by industry, as opposed to detailing how the goal is to be accomplished. The regulator is charged with defining the safety standards that must be met and then checking that industry has the management systems in place to ensure such standards are achieved.

2.2.1 Benefits

The advantage of performance-based regulation is that companies are given a relatively high degree of freedom in selecting solutions that fulfill the official requirements. Such an approach calls upon the industry to continually improve upon previous lessons learned to meet the specified goal. Without the requirement that the solution for a problem be met by complying with a prescriptive regulation, each company is encouraged to innovate and determine the best method of meeting the specified goal.

2.2.2 Challenges

Performance-based regulations may, however, suffer from the potential difficulty in interpreting the desired performance necessary to meet the goal defined in the regulations. A performance-based regulation may require more analysis and guidance documentation to determine that the specified goal is met. [3]

2.3 Regulation in Practice

Although a tendency exists to characterize a regulatory approach as either performance-based or prescriptive, there is no such thing as a purely performance-based or a purely prescriptive regulatory regime, although the regime in question can ultimately be classified as more performance-based or more prescriptive. The reality is that the most likely and preferred approach is often a blend of the two. Therefore, the choice of a more prescriptive or more performance-based regulation must be made based upon the strengths and weaknesses of each approach. In fashioning a regulatory regime, policy-makers cannot exclude the possibility of prescriptive or performance-based elements. The regulatory regime must be chosen based upon the needs of the regulated industry. [4]

3. U.S. Regulation of Offshore Process Safety

Legislation relating to offshore process safety is primarily dealt with in the Outer Continental Shelf Lands Act. [5] The Act itself does not limit the federal government's ability to regulate offshore process safety. Currently, this legislation has been utilized to provide the Bureau of Safety and Environmental Enforcement (BSEE) with the regulatory authority to oversee major accident prevention and mitigation. In response to the Macondo incident, the Department of the Interior (DOI), which is responsible for offshore safety, has changed both its structure and its regulations. The current regulations concerning offshore process safety as it relates to major accident prevention are the Safety and Environmental Management Systems (SEMS) regulations. [6]

3.1 Pre-Macondo Regulation

Prior to the Macondo incident, DOI charged the Minerals Management Service (MMS) with oversight of offshore safety. In addition to this duty, the MMS was also charged with responsibility for offshore leasing, revenue collection and auditing, permitting, and environmental protection. Such duties not only taxed the agency's resources, but also resulted in serious conflict of interest concerns. For instance, the billions of dollars at stake from offshore royalties occupied the agency's resources at the expense of offshore safety. [7]

Pre-Macondo regulation was largely concerned with a more prescriptive approach to oversight of offshore safety. The regulations concerned specific technical requirements such as pollution prevention, drilling operations, production requirements, and so forth. Such a regime led to a checklist approach to ensuring safety. In the early 1990s, however, there was interest at MMS in moving its regulatory regime towards a more performance-based approach due to the offshore sector's continuously evolving and highly technical needs. However, due to political, industrial, and other interests the movement was never successful. It was not until the Macondo incident that DOI overhauled the regulation of offshore safety. [8]

3.2 Post-Macondo Regulation: SEMS

Post-Macondo, DOI created BSEE to enforce its safety and environmental regulations. This was done in an effort to avoid the prior conflicts of interest that MMS experienced. [7] Furthermore, BSEE promulgated a host of new regulations addressing oil and gas offshore safety, including the Safety and Environmental Management Systems rule (SEMS). The final rule, issued in October 2010, incorporates by reference, and makes mandatory, API RP 75(3rd Edition). SEMS established requirements pertaining to 13 specific safety management elements, including hazard analysis, management of change, operating procedures, and training, among others. [6]

In September 2011, BSEE published a proposed amendment to the SEMS regulation with additional provisions; [9] informally called "SEMS II," the rule was finalized and published in April 2013. [10] The SEMS II rule provides additional requirements for stop work authority and ultimate work authority, employee participation in the development and implementation of SEMS programs, the reporting of unsafe working conditions, independent third-party audits of operators' SEMS programs, and conducting job safety analysis (JSA) for activities identified in an operator's SEMS program.

BSEE classifies the SEMS regulation as a more performance-based approach to the regulation of offshore safety. It recognizes that the offshore industry's constant evolution and highly technical nature would be best suited to a more performance-based regulatory approach. BSEE states that the SEMS regulation's performance-based approach focuses industry on the goal of risk minimization as opposed to creating a check-the-box type of approach common to prescriptive regulation. [11] [12] Certainly, in comparison with the previous pre-Macondo regulations, the SEMS regulation has moved BSEE toward a more performance-based framework.

However, while the SEMS regulation has made improvements in the management of offshore process safety, many stakeholders have called into question BSEE's classification of SEMS as a performance-based regulation. For instance, BSEE's own advisory committee, the Ocean Energy Safety Advisory Committee, has held that SEMS is too prescriptive. It stated that it feared such an approach would focus the regulated community on a culture of compliance instead of a culture of safety. [13] Various other stakeholders have raised similar concerns including Det Norske Veritas (DNV), Andrew Hopkins, the PEMBINA Institute, and the trade association, International Association of Drilling Contractors. [2] [3] [14] [15] Many have encouraged BSEE to look to international peers who embrace what they believe to be examples of more performance-based approaches to the regulation of offshore safety.

BSEE has stated that "legal, political, and operational environments vary from one country to another," and as such it is difficult to tailor the SEMS regulation to another country's regulatory framework. [10] Nevertheless, the CSB has taken an active interest in investigating the regulatory approaches taken by the United Kingdom, Norway, and Australia in an attempt to determine what benefits each regime could offer to offshore safety on the U.S. outer continental shelf.

4. International Approaches

4.1 The United Kingdom

Offshore safety in the United Kingdom is overseen by the U.K. Health and Safety Executive (HSE). The regulation of offshore safety in the U.K. was initially prescriptive in its approach. However, the impetus for the regulatory shift from prescriptive regulations to a more performance-based approach arose out of the 1988 Piper Alpha offshore platform fire and explosion that resulted in 167 fatalities in the North Sea. As a result of the incident, an inquiry was set up by the U.K. government to determine what could be learned. Importantly, the U.K. determined that any new offshore regulations must be risk-based and objective- or goal-setting in nature. This new regulatory framework is referred to as the safety case regime.

In essence, the safety case regime requires the operator of every installation to prepare an argument – a report that makes the case for safety – for the regulator that demonstrates that hazards have been identified, risks have been assessed according to a valid methodology, and risks have been reduced to as low as reasonably practicable (ALARP). The regulator must then either accept or reject the case. This highlights the importance of a regulator that has the competency to understand and enforce each respective safety case. The HSE actively pursues regulatory staff that collectively carries appropriate expertise.

The U.K. HSE provides a suite of helpful guidance about determining what constitutes ALARP. Reducing risk to ALARP is about weighing the process risk against the sacrifice needed to further reduce it. The decision weighs in favor of health and safety because the law presumes that the operator should implement the risk reduction measure. To avoid having to make this sacrifice, the duty-holder must be able to show that the proposed risk reduction technique would be grossly disproportionate to the benefits that would be achieved. The process is not simply a cost benefit analysis. Rather, duty-holders should adopt all measures except where they are ruled out because they involve grossly disproportionate sacrifices. [16]

The safety case report is intended to be a living and current document. The HSE intends the focus of the case be on continual improvement. This requires industry to consider technological, legal, and organizational changes encountered offshore. Further, the HSE holds that the performance-based nature of its safety case regulations provides the public with regulatory reassurance that high hazard risks are being properly managed. [1]

4.2 Norway

Norway oversees offshore safety through the Petroleum Safety Authority (PSA). The PSA administers five applicable set of regulations that range in nature from broad goal-based applications to more narrow requirements. This regulatory approach is more performance-based in that it requires few specific technical requirements. Rather, regulations concern management of risk through various management systems. [1] [17]

To assist operators in attaining these goal-based regulations, the PSA regularly publishes non-legally mandated guidelines and interpretations for each regulation. Examples and alternatives are provided for each operation to utilize in determining the choice best suited to its needs. In many cases, the guidelines and interpretations refer to standards. Furthermore, PSA interconnects each regulation to other pieces of legislation to provide operations with guidance in compliance and in understanding the reliance each safety element plays in relation to one another. [17]

Norway adopted a performance-based approach to offshore safety after it determined that a prescriptive approach was ineffective. This was brought to light in the face of major incidents encountered in Norwegian waters. [18] Norway also found that the maintenance required by prescriptive regulation was very high and severely taxed PSA's resources. Through the adoption of performance-based regulation the PSA has the ability to mandate the level of safety compliance, while at the same time allowing each offshore installation the opportunity to tailor the regulations to its particular needs. Furthermore, the duty to ensure that the regulatory goals are attained lies with the operator. Not only does this ensure that a compliance-based mentality is avoided, but it removes the demand upon the regulator. PSA utilizes a highly competent staff to ensure that the operator is effectively meeting the regulatory requirements.

4.3 Australia

The National Offshore Petroleum Safety and Environmental Management Authority (NOPSEMA) has the responsibility for regulating offshore safety in Australia. As opposed to the approach taken by the other regimes mentioned, Australia did not wait for its own offshore major accident before enacting legislation giving effect to the safety case regime. Prior to the adoption of the safety case regime in 1992, Australia also had primarily prescriptive regulation in place. The Australian regime can be classified as more performance-based regulation that focuses the operator on the duty of assuring that all operational risks are reduced to ALARP. [19]

Implementation of the safety case regime in Australia was not without its growing pains. Many issues were identified between 1992 and 2005 such as unclear boundaries, overlapping legislation, inconsistent application of the regulations, inadequate regulatory competency, and excessive legislation. In an effort to address these concerns, a simplified set of legislation compromising one act and four sets of regulations was promulgated, replacing many acts and hundreds of regulations. It is NOPSEMA's belief that this streamlined approach focuses the offshore industry on the key aspect of the safety case regime: continuous improvement. [1]

The legislation seeks to focus the operator on reducing its respective risks to ALARP. This is done by following a rigorous process of hazard identification and risk minimization using the appropriate mix of decision-making tools. It is the operator's duty to demonstrate that the selected tools are consistent with reducing risk to ALARP. The regulator verifies compliance with the ALARP standard via workforce consultation, investigation of complaints by the workforce, leading and laggings process safety indicators, compliance with best practice standards, and ultimately the regulatory staff's expertise.

5. Comparison

In an effort to provide the reader with clarity regarding the regulatory regimes discussed, the key elements of each country's regulations will be addressed. This section will also provide the reader with suggestions regarding the regulation of offshore safety in the U.S. These suggestions were provided to the CSB during its December 15, 2010 Public Hearing, through various outreach efforts, and in publications released in the aftermath of Macondo. The purpose of this section of the paper is twofold: (1) it provides the reader with a jumping off point for discussion about ways in which to further improve the regulation of offshore safety, and (2) it acts as a primer for the CSB's forthcoming regulatory analysis the Macondo incident.

5.1 Key Elements of Each Regime

5.1.1 United States

- Safety management system utilized
- The lessee or the owner or holder of operating rights is responsible for safety
- Duty to identify, address, and manage safety; no risk reduction target or goal
- Regulator is charged with ensuring compliance
- Management determines which employees participate and the level of participation in offshore safety
- Regulator is part of agency responsible for generating lease fees and royalties

5.1.2 United Kingdom

- Safety management system utilized
- Dutyholder is responsible for safety
    - Dutyholder is defined as one who create or has the greatest control over the risks
- General duty to reduce risk to ALARP
- Dutyholder is charged with ensuring compliance (Safety Case)
- Regulator is charged with verifying that the dutyholder's safety case for the prevention of major accidents adequately reduces risks to ALARP
- All employees have regulatory right to participate in offshore safety through workforce-elected safety representatives acknowledged by the regulator
- Regulator independent of revenue generating arm

5.1.3 Norway

- Safety management system utilized
- The obligated party is responsible for safety
    - The obligated party is defined as one with control and instruction authority
- General duty to reduce risk to the extent possible
- Obligated party is charged with ensuring compliance
- Regulator is charged with verifying that the obligated party's safety management program for the prevention of major accidents adequately reduces risk to the extent possible
- All employees have a regulatory right to participate in offshore safety through workforce-elected safety representatives acknowledged by the regulator
- Regulator independent of revenue generating arm

5.1.4 Australia

- Safety management system utilized
- The person(s) best positioned to control risk is responsible for safety
- General duty to reduce risk to ALARP
- Person(s) best position to control risk charged with ensuring compliance (Safety Case)
- Regulator is charged with verifying that the responsible party's safety management program adequately reduces risk to ALARP
- All employees have a regulatory right to participate in offshore safety through workforce-elected safety representatives acknowledged by the regulator
- Regulator independent of revenue generating arm

5.2 Suggestions for the U.S. Regulatory Framework

As mentioned above, various stakeholders have provided suggestions for improving the current U.S. approach to regulating offshore process safety. The reader is provided with a few of these suggestions in the hopes of facilitating a discussion on how to improve offshore process safety.

The quotes that follow were made in the aftermath of the Macondo incident and pursuant to changes made by BSEE.

5.2.1 National Commission on the BP Deepwater Horizon Oil Spill and Offshore Drilling

"The Department of the Interior should develop a proactive, risk-based performance approach specific to individual facilities, operations and environments, similar to the 'safety case' approach in the North Sea." [20]

5.2.2 Ocean Energy Safety Advisory Committee

"SEMS should be performance based and adapted to the needs and requirements of the business and the operating systems." [13]

5.2.3 DNV

"The current safety regime for the U.S. Gulf of Mexico is largely a prescriptive regulation with no requirement for safety cases to be performed. . . . DNV believes that an offshore safety regime based on a performance-based regulation requiring safety cases including risk assessments supplemented by required or recommended specific prescriptive regulation for selected areas is the most effective regime model." [3]

5.2.4 International Association of Drilling Contractors

"We urge BSEE to consider a wholesale re-write of [SEMS], . . . in order that [it] be made more goal-setting and less prescriptive." [15]

5.2.5 The PEMBINA Institute

"The new SEMS requirement, while leaving primary responsibility for SEMS implementation and oversight to the operator, is itself relatively prescriptive, mandating in detail the elements of an operator's SEMS program, rather than simply setting the objectives of such programs and leaving details to operators." [2]

5.2.6 Andrew Hopkins

"It is a depressing commentary on the US political system that the shock of the Macondo disaster has not been sufficient to shift the political balance in favour of a safety case regime." [14]

5.2.7 Shell Oil

"Since 2002 we've used a safety case globally in our operations for oil well drilling." [1]

6. Conclusion

The tragic consequences of the Macondo incident provide the offshore community with a multitude of opportunities to assess and improve upon the management of safety. In the face of many pervasive problems with offshore safety, which includes technical and organizational issues, an assessment of the regulation of safety is needed. Through an in-depth comparison of how process safety is regulated internationally, the U.S. can seek to improve upon its own regulations by adopting the aspects of other frameworks that drive continual improvement.

Whatever regulatory approach taken, it must be tailored to the particular industry's needs. A more prescriptive approach is better suited for industries where best practices can be clearly defined, there is little need or potential for innovation, and where deviation from requirements could create unacceptable risks to the environment or human health. A more performance-based approach is better suited to industries with non-standard operations, and complex and constantly changing needs. Indeed, it is not feasible to develop a set of comprehensive prescriptive regulations for a complex industry. Some of the more advanced and active countries participating in offshore oil and gas operations tend to favor the adoption of a more performance-based regulatory approach. Furthermore, many within the industry have advocated for the U.S. to adopt a similar approach as the offshore industry becomes ever more global, and offshore operations and technologies continue to advance and evolve.

The CSB is currently investigating ways to further improve the regulation of safety in U.S. oil and gas operations, both on and offshore. Public hearings, outreach efforts to stakeholders, and research have provided many insights into the issue. Papers such as this one arm the public, government, workforce, and industry with information on the various qualities of existing regulatory regimes in order to promote discussion on how the U.S. offshore oil and gas regulatory framework could be strengthened to prevent future catastrophic accidents. The CSB plans to release a report on this topic as it relates to the Macondo incident in 2014. Through these efforts the CSB seeks to elevate this important discussion and ultimately focus the regulator, industry, and the workforce on the most robust safety management framework for continuous improvement and risk reduction.

7. References

[1] U.S. Chemical Safety and Hazard Investigation Board, Regulatory Approaches to Offshore Oil and Gas Safety, Dec 15, 2010, http://www.csb.gov/assets/1/19/Transcript_of_Public_Meeting_12_15_2010.pdf (accessed Jan 24, 2014).

[2] The PEMBINA Institute, Comparing the Offshore Drilling Regulatory Regimes of the Canadian Arctic, the U.S., the U.K., Greenland and Norway, June 2011, http://www.pembina.org/pub/2227 (accessed Jan 24, 2014). The PEMBINA Institute is a Canadian organization seeking to advance clean energy solutions through innovative research, education, consulting and advocacy. For more information see: http://www.pembina.org/about/about-pembina (accessed Jan 24, 2014).

[3] Pitblado, Robin, et. al., Key Aspects of an Effective US Offshore Safety Regime, July 22, 2010.

[4] U.S. Nuclear Regulatory Commission, Guidance for Performance-Based Regulation: NUREG/BR-0303, Dec 2002.

[5] Outer Continental Shelf Lands Act, 43 U.S.C. §§ 1331–1356a.

[6] 30 C.F.R. § 250 Subpart S (2014).

[7] U.S. Department of the Interior, Salazar Divides MMS's Three Conflicting Missions, May 19, 2010, http://www.doi.gov/news/pressreleases/Salazar-Divides-MMSs-Three-Conflicting-Missions.cfm, (accessed Jan 24, 2014).

[8] *See* National Commission on the BP Deepwater Horizon Oil Spill and Offshore Drilling, Deep Water: The Gulf Oil Disaster and the Future of Offshore Drilling, Jan 2011; *see also* Marine Board of the National Research Council Committee on Alternatives for Inspection of Outer Continental Shelf Operations, Alternatives for Inspecting Outer Continental Shelf Operations, 1990, http://www.nap.edu/openbook.php?record_id=1517&page=1, (accessed Jan 24, 2014); *see also* T.A.F. Powell, US Voluntary SEMP Initiative: Holy Grail or Poisoned Chalice?, Offshore Technology Conference [OTC] Paper 8111, May 1996.

[9] Proposed Rule, Oil and Gas and Sulphur Operations in the Outer Continental Shelf--Revisions to Safety and Environmental Management Systems, 76 Fed. Reg. 56,684 (Sept 14, 2011).

[10] Final Rule, Oil and Gas and Sulphur Operations in the Outer Continental Shelf—Revisions to Safety and Environmental Management Systems, 78 Fed. Reg. 20,423 (Apr 5, 2013).

[11] Final Rule, Oil and Gas and Sulphur Operations in the Outer Continental Shelf—Safety and Environmental Management Systems, 75 Fed. Reg. 63,610 (Oct 15, 2010).

[12] Dlouhy, Jennifer A., Offshore enforcement chief outlines approach to safety, Dec 18, 2012, http://fuelfix.com/blog/2012/12/18/offshore-enforcement-chief-outlines-approach-to-safety/ (accessed Jan 24, 2014).

[13] Ocean Energy Safety Advisory Committee, January 9-10 Committee Recommendations, Jan 25, 2013, http://www.bsee.gov/uploadedFiles/BSEE/About_BSEE/Public_Engagement/Ocean_Energy_Safety_Advisory_Committee/OESC%20Recommendations%20January%202013%20Meeting%20Chairman%20Letter%20to%20BSEE%20012513.pdf (accessed Jan 24, 2014).

[14] Hopkins, Andrew, Disastrous Decisions: The Human and Organisational Causes of the Gulf of Mexico Blowout, 2012. Emeritus Professor Hopkins is an Associate Member of the National Research Centre for OHS Regulation and is an adjunct Professor with the School of Sociology in the College of Arts and Social Sciences at The Australian National University. For more information see: http://regnet.anu.edu.au/people/emeritus-andrew-hopkins (accessed Jan 24, 2014).

[15] International Association of Drilling Contractors, Revisions to Safety and Environmental Management Systems (SEMS), RIN 1010-AD73, Nov 11, 2011. The International Association of Drilling Contractors represents the worldwide oil and gas drilling industry. For more information see: http://www.iadc.org/about-iadc/ (accessed Jan 24, 2014).

[16] U.K. HSE, ALARP "at a glance," http://www.hse.gov.uk/risk/theory/alarpglance.htm (last visited Jan 24, 2014).

[17] Norway Petroleum Safety Authority, Regulations, 2014, http://www.ptil.no/regulations/category873.html, (accessed Jan 24, 2014).

[18] Norway Petroleum Safety Authority, From prescription to performance in petroleum supervision, Dec 3, 2010, http://www.ptil.no/news/from-prescription-to-performance-in-petroleum-supervision-article6696-878.html, (accessed Jan 24, 2014).

[19] NOPSEMA, ALARP Guidance Note: N-04300-GN0166 Revision 4, Dec 2012, http://www.nopsema.gov.au/assets/Guidance-notes/N-04300-GN0166-ALARP.pdf, (accessed Jan 24, 2014).

[20] National Commission on the BP Deepwater Horizon Oil Spill and Offshore Drilling, Deep Water: The Gulf Oil Disaster and the Future of Offshore Drilling, Jan 2011.

Additional references not cited:

[21] Hopkins, Andrew, The Meaning of "Safety Case," Australian National University, Feb 2013.

[22] U.S. Chemical Safety and Hazard Investigation Board, Working Papers on the Safety Case Regulatory Model and its Attributes, Jan 7, 2014, http://www.csb.gov/working-papers-on-the-safety-case-regulatory-model-and-its-attributes/ (accessed Jan 24, 2014).

10TH GLOBAL CONGRESS ON PROCESS SAFETY

Implementation of Design Best Practices for Upstream Process Safety

Sandipan Laskar
Sr. Process Safety Manager
Clean Harbors Inc.
2027 Independence PKWY S
La Porte, TX 77571
laskar.sandipan@cleanharbors.com

Prepared for Presentation at
American Institute of Chemical Engineers
2014 Spring Meeting
10th Global Congress on Process Safety
New Orleans, LA
March 30 – April 2, 2014

Implementation of Design Best Practices for Upstream Process Safety

Sandipan Laskar
Sr. Process Safety Manager
Clean Harbors Inc.
2027 Independence PKWY S
La Porte, TX 77571
laskar.sandipan@cleanharbors.com

Keywords: Safety and Environmental Management System (SEMS), Upstream Process Safety.

Abstract

Although not all upstream facilities have been historically obligated to meet the requirements of the federal OSHA PSM standard (29 CFR 1910.119), utilizing this existing set of technical, documentation, and procedural practices can be expected to achieve the same results as in the downstream, chemical, and petrochemical industries. Namely, to ensure the safe operation of a facility, various hazards associated with the facility process and the plant equipment need to be addressed in a systematic manner.

As a result of the Deepwater Horizon incident on April 20, 2010, the Department of the Interior adopted the Safety and Environmental Management Systems (SEMS) regulation (30 CFR 250). This regulatory program, which imposes a process safety management-like program on oil and gas facilities operating on the Outer Continental Shelf (OCS), replaces a voluntary program that had existed since 1993.

Those working in the oil and gas industry are reminded with depressing regularity of the importance of process safety management, as post-incident investigation invariably points to one or more deficiencies related to process safety as root causes. This paper challenges the predominant notion that upstream facilities can be safely operated without following methods similar to those dictated by OSHA based on technical comparisons of pressure relief system design findings for upstream (petroleum) and downstream (chemical process) facilities. An adequate pressure relief system plays an important role in safe facility operation both through avoidance of catastrophic events and through prevention of incident escalation.

Similarities in relief system design issues for multiple process units of representative upstream and downstream production facilities are presented in a comparative case study. Alignments and points of divergence in typical practice between the example facilities are discussed. Significant similarity in the nature and types of the pressure relief systems deficiencies between upstream and downstream facilities is concluded.

The information presented helps upstream asset owners better understand standard practice for OSHA regulated facilities, align the corporate approach to safety, especially for vertically integrated owners, and demonstrate benefits achieved through proper design and documentation.

1. Introduction

Process safety has become one of the inevitable elements in keeping businesses in various industries sustainable. Various industries have suffered painful lessons from incidents and have taken steps to prevent similar tragedies from re-occurring. The Bhopal incident in 1984 and the more recent Deepwater Horizon incident are few examples which show the catastrophic impact of process safety related incidents. Although eliminating risk entirely is not feasible, especially in handling highly hazardous chemicals, recent progress in process safety has improved safety performance through recognizing the risk and implementing safety measures to prevent or mitigate catastrophic consequences.

Probably the world's worst chemical disaster, the Bhopal MIC release in 1984, triggered new comprehensive regulations in the US chemical industry. Following the Bhopal incident, the Emergency Planning and Community Right-to-Know Act (EPCRA) of 1986 was passed by the U.S. Congress. This act promotes emergency planning for potential chemical accidents and provides communities with proper information on potential hazards imposed by the chemical processes. The development of formal Safety Management Systems for offshore facilities can be said to have started with the Piper Alpha catastrophe that occurred in 1988. Piper Alpha is often viewed as being the starting point for the development of Safety Management Systems for offshore operations.

In the course of developing these safety systems, despite different approaches to offshore safety management taken by organizations, there are many similarities between SEMS and Safety Cases (and onshore Process Safety Management programs). Fundamentally, they all result in the development of a risk-based Safety Management Systems, and they make extensive reference to prescriptive industry standards.

Based on the fundamentally different criteria for risk, two different approaches can be implemented when developing the requirements for regulating process safety elements. Performance based regulations specify desired risk level as a goal. Operators must prove that the operating design and procedures meet the goal with detailed explanation and quantified risk assessments. Performance based regulations require more effort in evaluating compliance; it can provide more flexible approaches in accommodating latest technologies under various operating conditions. Prescriptive based regulations provide detailed requirements based on certain engineering best practices. Checking for compliance for the prescriptive based regulation is straightforward compared to performance-based regulation, as detailed requirements have already been set. The new Safety and Environmental Management System (SEMS) requirement for OCS oil and gas operation in the US is a prescriptive based regulation [1].

2. Regulations

Safety and Environmental Management Program (SEMP)

For offshore facilities, a more holistic approach applying to both production and drilling and without a threshold volume was published originally in 1998 by the API, with the most recent update in 2004 (API 2004). The recommended practice 75 (API RP 75) establishes methods to develop a Safety and Environmental Management Program (SEMP) covering 12 target elements, which addresses all the major aspects of an OSHA PSM program. Although the MMS and industry groups ran a number of audit surveys during the initial implementation period of API RP 75, recent compliance status data are not publicly available [1][3].

The MMS published an Advance Notice of Proposed Rulemaking (ANPR) in the May 2006 US Federal Register related to the creation of an enforceable PSM/SEMP, and provided a report in 2009 on the comments and data gathered during the voluntary PSM/SEMP period (MMS 2009). Data analysis was performed on a number of MMS incident reports, noncompliance notices, and spill reports, showing a preponderance of incidents resulting from failures in one or more of four main program areas.

Historically the MMS has not been regarded as a strong enforcement agency, though with the reorganization announced by the Interior Secretary in May 2010 (Straub 2010) splitting the field development, enforcement, and revenue functions of the former MMS body, in combination with the promulgation of an enforceable PSM/SEMP and the recent major catastrophe in the Gulf of Mexico, one might have expected a more stringent regulatory environment with greater emphasis on design review and prevention as well as increased penalties and fines should incidents occur. Following a flurry of activity in the second half of 2010, the SEMP as described in API RP 75 became mandatory [4]

API RP 75 is basically a performance based and non prescriptive standard does not provide much detail to do with the implementation of the elements. API RP 75 does reference a large number of other standards. Most of these are from API itself but some are from other professional organization such as the ASME (American Society of Mechanical Engineers) [5].

Process Safety Management (PSM)

Process Safety Management (PSM) refers to a set of inter-related approaches to manage hazards associated with the process industries and is intended to reduce the frequency and severity of incidents resulting from releases of chemicals and other energy sources. These standards are composed of organizational and operational procedures, design guidance, audit programs, and a host of other methods. The different elements of the PSM and SEMP program, is as shown below [2].

These guidelines are performance based and as such do not provide explicit techniques to achieve the goals of the program. For instance, with respect to compilation of process safety information described in subpart d, the standard requires the compilation of a number of relevant data types in a current and accurate state, but does not specify format or style of presentation, define these data types, or describe methods to validate content. Generally the OSHA guidelines rely on recognized and generally accepted good engineering practice (RAGAGEP), but allows companies to utilize alternate methods as long as they can prove that performance resulting from that method meets the standard. The specifications and techniques that form RAGAGEP are generally understood to mean recommended practices from the American Petroleum Institute (API), codes from the American Society of Mechanical Engineers (ASME), standards from the International Standards Organization (ISO), certification agency standards (DNV, Lloyd's, ABS), and other sources in regular use.

Table 1- Comparison between OSHA PSM and upstream SEMP

OSHA's PSM	SEMP
1. Employee Participation	1. Safety and Environmental Information
2. Process Safety Information	2. Hazard Analysis
3. Process Hazard Analysis	3. Operating Procedures
4. Operating Procedures	4. Training
5. Training	5. Pre-Startup Review
6. Contractors	6. Assurance of Quality and Mechanical Integrity of Equipment
7. Pre-startup Safety Review	7. Safe Work Practices
8. Mechanical Integrity	8. Management of Change
9. Hot Work	9. Investigation of Incidents
10. Management of Change	10. Emergency Response and Control
11. Incident Investigation	11. Audit of Safety and Environmental Management Program Elements
12. Emergency Planning and Response	12. Records and Documentation
13. Compliance Audits	
14. Trade Secrets	

The OSHA PSM guidelines are enforceable throughout the United States and specific US holdings for facilities having a threshold volume of hazardous chemicals, but are not applicable where another federal agency has occupational safety and health jurisdiction, including in federal waters controlled by Bureau of Ocean Management Regulation and Enforcement (BOEMRE, formerly Minerals Management Service (MMS)). Oil and gas drilling and servicing operations and normally unmanned remote facilities are specifically excluded from the standard, but typical facilities in the upstream industry such as production facilities, gas oil separation plants, and gas plants are covered if the threshold chemical volumes, typically hydrocarbons, are present.

3. Case Study

The main function of a pressure relieving system is to provide a positive means of protection for equipments from over-pressurization above limits permitted by the relevant codes and to provide a safe relief path. Identification of the causes of overpressure, quantification of relieving rates required complying with codes, sizing of relief devices and the design of relief disposal system is a complex task. Use of systematic method for the identification of contingencies, together with a set of project specific guidelines of the governing standards and codes, provide the basis for sound relief system design. Along with proper design, proper documentation of the above steps is of utmost importance. This paper challenges the predominant notion that upstream facilities can be safely operated without following methods similar to those dictated by OSHA based on technical comparisons of pressure relief system design findings for upstream and downstream facilities. An adequate pressure relief system plays an important role in safe facility operation both through avoidance of catastrophic events and through prevention of incident escalation. To support the comparison, pressure relief systems verification findings from 60 chemical process industries, i.e., downstream, process units and 35 offshore and upstream production facilities were analyzed.

The downstream process units contain 9,868 pieces of equipment and 4,897 pressure relief devices. The upstream process units contain 3,491 pieces of equipment and 2,438 pressure relief devices. The equipment to pressure relief devices ratio of the evaluated units was 2.0 for the downstream and 1.4 for the upstream units.

The ASME Boiler and Pressure Vessel Code Section VIII Division 1 (ASME VIII Division I) requires that each equipment covered by this code has overpressure protection to prevent loss of containment. The consequences of not meeting this requirement may jeopardize the safety of the facility. In performing these pressure relief systems verification case studies, a best practice equipment-based approach has been utilized. This approach covers each piece of equipment in the operating facility regardless of the presence of a relief device on equipment [6].

The scope of the pressure relief system verification was to identify the deviations from RAGAGEP on existing process installations in four categories, as follows:

1. No pressure relief device;

This category includes all pieces of equipment where an overpressure scenario was concluded to be applicable, but no dedicated pressure relief device was installed.

2. Undersized pressure relief device;

This category includes all pieces of equipment where a dedicated pressure relief device is installed; however, it was concluded through a detailed pressure relief analysis that the installed device was not adequate to provide protection for all identified overpressure scenarios.

3. Improper pressure relief device installation;

This category includes all installed relief valves that are associated with hydraulic deficiencies that can be attributed to the pressure relief device inlet and outlet piping design and installation. Excessive inlet and outlet piping frictional losses that did not meet API RP 520 (API 2008) and ASME VIII Division I (ASME 2010) were added to this category. In addition, pressure relief device set pressure issues and installations with the potential to block the relief path were included in this category.

4. Documentation Issues.

This category includes instances where the relief device datasheet have missing information. The documentation issue involves issues with relief valve sizing errors, scenario identification and improper selection of design codes.

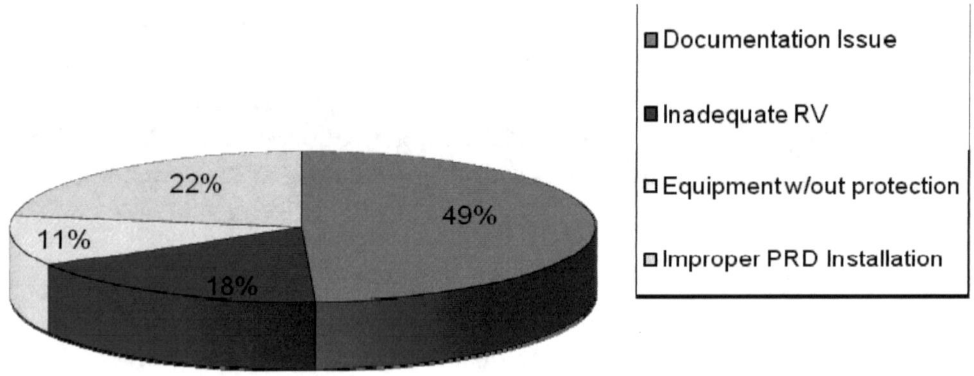

Figure 1: Pressure relief systems deficiencies distribution in evaluated upstream process units

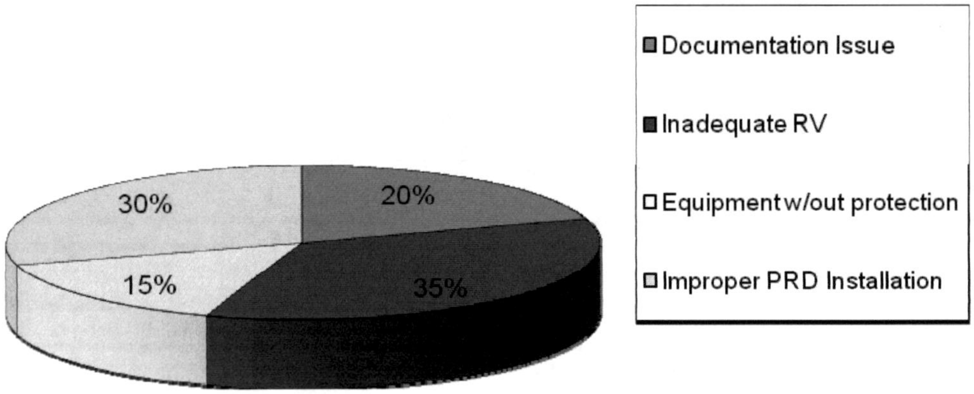

Figure 2: Pressure relief systems deficiencies distribution in evaluated downstream process units

Comparing the downstream and upstream process unit deficiencies indicates similarities in the nature of these findings. It is evident that the documentation concerns account for the largest portion of the installed pressure relief devices deficiencies for upstream facilities. The majority of the documentation concerns are related to incomplete relief valve datasheet information for upstream as well downstream facilities. Information regarding sizing basis and applicable RAGAGEP is also one of the major reasons for documentation discrepancies. However, undersized relief device deficiencies are higher within the downstream facilities, possibly due to the number and complexity of the downstream facilities overpressure scenarios. A closer look at the documentation issue suggested that missing or outdated information on the relief device datasheet is the major concern for the upstream facilities. When compared with the downstream facilities, the numbers of issues are less suggesting that the documentation methodology used in the downstream relief device documentation well established.

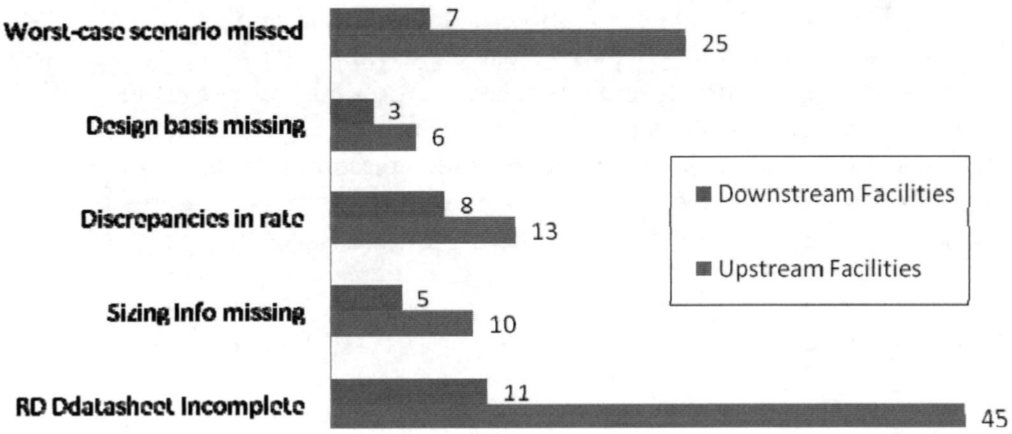

Figure 3: Pressure relief devices documentation concerns

4. Results

A total of 3,120 and 854 deficiencies were identified for the evaluated chemical process units and upstream units, respectively. The study indicates that approximately 63% of the installed pressure relief devices in the evaluated downstream facilities were associated with at least one of the identified deficiencies categories. That percentage dropped to 35% in the upstream facilities. Downstream refining process units are generally associated with a larger number of overpressure scenarios and several pieces of equipment may be protected by a single relief valve. Such arrangements may cause the number of installed relief devices deficiencies to be higher in comparison to upstream facilities, which tend to have less process complexity. Additionally, downstream facilities tend to be expanded more frequently than upstream facilities, which typically experience design flow rates during the plateau period early on, with subsequent decline. During these process modifications, analysis of relief system adequacy is often overlooked or not properly performed. A closer look in the documentation concern illustrates that a majority of the relief device datasheet had missing or misleading information.

5. Conclusion

The offshore industry is now challenged to improve safety performance in OCS operations with the new SEMS regulatory standard. Whereas none of the elements are new to current OCS operations, the requirements are now mandatory for all offshore operations. The SEMP described in API RP 75 mirrors the requirements of the well established OSHA PSM program. While it is true that application of the PSM program to onshore facilities over the past 15 years or so has not reduced the incident rate to zero, it is probable that numerous incidents with potential for catastrophic results were either mitigated or avoided entirely. It is clear that both PSM and SEMP rely in large part on vigorous enforcement and identification of issues before incidents occur.

Comparing the data presented, gathered from a variety of onshore and offshore facilities and focused on one PSM/SEMP element, it can be concluded that similar issues exist in upstream and downstream facilities. Although the evaluated process units did not necessarily share the same level of complexity, common findings related to pressure relief system verification should debunk any notion that upstream and downstream facilities differ significantly. This demonstrates that offshore facilities can improve the margin of safety through application of proven process safety management practices. Again, it makes sense to apply tools and practices already developed in the onshore industry to address similar issues offshore.

6. References

[1] *API RP 75, Recommended Practice for Development of a Safety and Environmental Management Program for Offshore Operations and Facilities*, third edition. 2004. Washington, DC: API

[2] *Process Safety Management of Highly Hazardous Chemicals*, U.S. Occupational Safety and Health Administration, 29 CFR 1910.119, Federal register, 57 (36) (Feb 24, 1992).

[3] Straub, N. 2010. Interior Unveils Plan to Split MMS into 3 Agencies. *The New York Times*, May 20, 2010.

[4] Bureau of Ocean Energy Management, Regulation and Enforcement Implementation (BOEMRE), *Oil and Gas and Sulphur Operations in the Outer Continental Shelf--Safety and Environmental Management Systems; Final Rule*, 30 CFR Part 250, Volume 74, Number 199, October 15, 2010.

[5] Bureau of Ocean Energy Management, Regulation and Enforcement Implementation (BOEMRE), *Survey for API RP 75*, September 17, 2010

[6] *API Std 520 – Part I, Standard for Sizing, Selection, and Installation of Pressure-Relieving Devices in Refineries*, eighth edition. 2008. Washington, DC: API.

352260 Monetising Gas Hydrates: New Mode for Natural Gas Storage and Transportation Against LNG Storage and Transportation

Abstract:

Gas Hydrates are considered in the petroleum industry e.g. natural gas transportation through pipelines, during natural gas production from the crude oil wells. But this paper focuses on converting the nuisance of gas hydrates to an attractive means of transportation. New Technology development has been focusing on using gas hydrates as a way of converting gas to solids.

The Natural Gas Hydrate (NGH) is a clatherate compound in icy state and stabilised around -20 degree Celsius under atmosphere. This paper summarises about the cost effectiveness of the NGH value chain against the LNG (Liquefied Natural Gas) value chain and also in this paper entire processes and operations to create NGH pellets are designed and discussed. The calculations show that this process has a high initial plant cost but on a long term this process of NGH production is cost effective compared to LNG liquefaction process. Moreover, it is easy to handle the NGH at -20 to -30 degree Celsius compared to LNG at around -160 degree Celsius.

Thus, this unique method of transporting the natural gas in bulk solid form leads to reduction of the total investment cost for the entire value chain. Moreover, NGH is expected to be a plausible solution in monetising small to medium gas fields which are considered to be difficult to be developed economically using LNG.

352647 Towards the Development of a Control-Relevant Model of the Hydraulic Fracturing Process to Investigate Various Control Strategies

Abstract:

Hydraulic fracturing or 'fracking' is an increasingly important technique for developing energy resources across both the U.S. and the world, in which high pressure fluid is injected into the reservoir to crack the rock formation and provide highly conductive channels for oil and gas transport back to the wellbore. During this process, the underground environment is characterized by complex fluid dynamics, geo-mechanics, and multiphase flow phenomena. Surface equipment exerts control over the process from between hundreds of feet to several miles from the moving front of the fracture(s) below ground. The fluid pumped underground is a slurry consisting of water, proppant (often a special type of sand) to keep the created fractures open after the process finishes, and chemical additives such as cross-linkers, acids, and surfactants to maintain fluid viscosity, reduce corrosion, and perform other duties. During the fracturing process, pumping rate and fluid properties (viscosity and the concentrations of proppant and other additives) can be varied at the surface. The main objectives are to keep the fracture growth within a predefined region of the reservoir open by distributing the proppant evenly throughout the fracture so that oil and gas can flow after the fracturing fluids have leaked into the formation. Complications can include excessive growth of the fracture external to the target region that results in the proppant being spread so thinly that it is ineffective to keep the fracture open. Conversely, if a high concentration of proppant is added to the fracturing fluid mix too early in the process the condition called 'screen-out' can occur wherein the proppant enters the advancing fracture front and bridges the tip preventing further fracture growth and causing rapid pressure buildup in the system. A related issue is proppant settling, that is a large sand bank or dune forms near the entrance of the fracture at the wellbore resulting in only a small open region around the wellbore after the fluid leaks off. These three potential proppant-related challenges may have a large negative impact on oil well production.

Applying process control techniques to regulate the hydraulic fracturing process, via setting optimal pumping schedules, is a challenging problem not only due to the remoteness of the control action, but also due to the limited feedback information available because of the sensing system associated with the bottom-hole situation. To better understand these challenges, a mathematical model of the physics of the fracturing system is developed to determine the effect of different pumping schedules on the final fracture conductivity, the most widely used calculated performance metric. The physics-based model incorporates equations governing fracture propagation, fluid dynamics, proppant transport, and bank formation. The resulting system of nonlinear, partial differential equations is solved numerically using a moving grid method to track the fracture growth and particle tracking to predict the proppant movement and settling. Additionally, the 'shut-in' process, during which pumping is stopped and the rock formation walls are allowed to close as the fluid leaks off is a part of this model development. The final geometry of the channel available to oil and gas flow is predicted. Using this model, improved pumping schedules can be developed and tested to achieve near-optimal fracture conductivity.

352284 Shale Gas as a Potential Source of Unconventional Reserve for Crude Oil

Abstract:

With a decrease in the conventional reserves available for the recovery of oil, the alternative fossil fuel extraction industry is on a growth. As the development of unconventional oil and gas resources has been progressed, the heterogeneity and complexity of shale gas and tight oil reservoirs have become apparent. In the U.S. primarily shale gas and tight oil is expected to continue growing dramatically in coming decades. According to the U.S. Energy Information Administration, an assessment of 137 shale formations are identified in 41 countries other than those identified in U.S. At the local level, though the environmental impacts of shale gas and tight oil development may turn out to be significant.

Shale gas and tight oil are unconventional sources of petroleum that are found within the reservoirs with a very low permeability. The oil and gas contained within these reservoir rocks typically will not flow to the wellbore at the economic rates without assistance from technologically advanced drilling and completion processes. Usually, horizontal drilling coupled with multi stage fracturing is used to access these difficult to produce reservoirs. Unconventional natural gas and oil deposits are difficult to characterize overall, but in general are often lower in resource concentration, more dispersed over large area, and require well stimulation or some other extraction or conversion technologies.

352037 Models Describing the Thermal Recovery as a Productive Option for the Heavy Oil Reservoirs

Abstract:

The crude oil requirement is increasing relatively at a higher rate in comparison to the amount of new crude reserves being identified. To meet the requirement, the recovery from the available crude reserves should be maximized. After the primary recovery, there is a still a good proportion of crude is still left in the reserve. There is a need for the secondary and tertiary methods for the recovery of the left out crude.

Thermal methods are one of the successful techniques which can be implemented to enhance the recovery. I have been tested since 1950's, and they are the most advanced among all the enhanced oil recovery methods available, as far as field experience and technology are concerned. The best utilization of thermal methods can be in the reserve left with heavy oils (of 10-20° API) and tar sands (≤10° API). Thermal methods provide a source of energy in the reservoir resulting in the large reduction in the viscosity of the heavy oil left. Decrease in the viscosity will result in a higher mobility ratio.

Thermal methods have been highly successful in Canada, USA, Venezuela, Indonesia and other countries. The general classification of Thermal enhanced oil recovery can be broadly classified in terms of hot water, steam, in-situ gasification or electrical heating. Among these methods the steam based methods are more successful commercially than others. In the present study, we will be discussing about the various models used to understand the behaviours of the reservoir in response to the thermal enhanced oil recovery methods. We will be discussing about what are the type of simulators being used to identify the behaviour.

Fire flooding as a budding source of enhanced oil recovery

- by KHUSHBOO SINGH (f2011409@pilani.bits-pilani.ac.in)

Fire flooding is a thermal method of oil recovery in which compressed air is injected into an injection well and the petroleum in the reservoir surrounding the injection well is ignited to establish a combustion zone or fire front .This fire front propagates in a radially expanding manner away from the injection well. The heat reduces the viscosity of oil, cracks some higher molecular weight hydrocarbons into smaller ones, vaporizes some lighter hydrocarbons to help miscibly displace oil and also produces steam that may steam distill the trapped oil .Thus, along with decreasing the mobility ratio, it also produces improved oil. Also, the gases produced due to combustion and nitrogen (in air) displace the oil towards reservoir away from fire front again enhancing production .Other methods of EOR cannot be used for highly viscous oils whereas thermal recovery or say, fire flooding can be used for both high and less viscous oils. Fire flooding also offers many advantages over the other method of thermal oil recovery i.e. steam flooding. Lot of fresh water and capital is used to produce steam for steam flooding whereas in this there is just the need to push the compressed air and ignite the oil thus, lowering the cost of production .Also, it has the potential to operate in wide variety of reservoirs that other methods. For example, lower in pressure (as it lowers the viscosity (improved mobility)as well as produces gases that supply pressure), containing more shale , lower in quality and thinner (as it requires less volume of air for producing the same amount of oil than other methods which employ gases like carbon dioxide method of oil recovery)and deeper (heat losses occur in steam flooding).But it also has some limitations like it can be applied only to reservoirs of high permeability , high porosity and oil saturation and has poor sweep efficiency in thick formations .Also, if excessive coke is deposited the rate of advancement of combustion zone will be slow and more air will be required. It also produces flue gases which cause environmental problems. Along with this it also faces operational problems like corrosion due to low pH hot water, serious oil water emulsions, and increased sand productions, deposition of carbon or wax and pipe failures to high temperatures.

Now however, oxygen enriched air or pure oxygen is being used as the combustion supporting medium for which about one-fifth as much volume is required for the injection process as compared to compressed air since nitrogen which accounts for about 80% of the volume of the compressed air is eliminated. Thus, the use of the pure oxygen in the process instead of air not only drastically reduces the volume of the combustion supporting medium required but also provides a decrease in the fluid volume following combustion which generally increases the injectivity of the reservoir matrix in the combustion zone .Also, it produces carbon dioxide which dissolves in oil and reduces its viscosity. Thus, one of the major limitations of fire flooding i.e. it requirement of high permeability in reservoirs is eliminated in this way .After an initial level of in situ combustion the production well is partially choked or shut in until the bottom hole pressure increases to a substantial fraction of the injection pressure .The production well can then be opened to a lower back pressure level which results in an immediate acceleration of fluid flow under the increases oil pressure gradient and leads to enhanced recovery. Thus, pressurization and producing cycles can be repeated using intervals found to be most effective for the particular system. Also, water or steam or mixtures of steam and carbon dioxide (or low molecular weight C3-C8 hydrocarbons) can be injected intermittently or subsequent to injection of combustion supporting medium to enhance production. Further, relatively heavy viscous crude oil from a subterranean stratum can be produced by the use of in situ combustion. In conventional direct drive or forward underground in-situ combustion, when applied to certain oil-bearing strata, particularly those containing a heavy viscous crude oil, it becomes impossible to continue to support an expanding combustion zone because of the formation of a viscous liquid bank of oil in advance of the combustion zone. This viscous liquid bank, substantially completely plugs the stratum and prevents continued injection of air to feed the combustion zone .So, now the oil is ignited in the stratum to be produced around a single well penetrating same and air is injected through the well to expand and move the ignited area (combustion zone) into the stratum away from the well whereby a liquid bank of viscous oil builds up in the stratum beyond the combustion zone so as to substantially plug the stratum and build up substantial pressure above the stratum pressure at the time of initiation of the process, and thereafter terminating the injection of air and opening the well to production so that the fluid pressure in the stratum beyond the combustion zone forces liquid oil from the liquid bank back into the combustion zone and the hot burned-over area behind the combustion zone so as to force the upgraded oil into the well from which it is recovered by conventional methods. After a substantial period of production during which the pressure-build up in the stratum during the injection period forces oil in upgraded condition into the well, the stratum is again ignited adjacent the well to establish a combustion zone which feeds on the oil filling the pores of the stratum at the end of the first production phase. In the event the stratum is below combustion supporting temperature, the ignition is effected in conventional manner as by heating with an electric igniter, an ice charcoal pack, a gas fired heater, or similar device and injecting air into the hot stratum. Also, recovery can be improved using water containing surfactant and alternate slugs of air during forward combustion. In an underground in situ combustion process for the recovery of oil from formations having high permeabilities, after combustion has been initiated, water containing a surfactant and air can be injected under conditions such that the water upon contact with the hot formation vaporizes, creating slugs of condensable foam in an amount sufficient to generate a significant resistance factor of at least 5, which improves volumetric sweep efficiency by retarding the mobility of the steam and subsequent slugs of air. Air and water, including a surfactant, can be injected simultaneously as a mixture or as alternate slugs of water containing a surfactant and air. It also provide a method for temporarily or partially plugging the more permeable strata and a process for producing a well by selectively blocking and plugging highly permeable sections of subterranean oil-bearing formations with a self-destructive foam generated in situ exhibiting a significant resistance factor. Thus, as we saw in situ combustion can be applied to a wide array of reservoirs whether it be thin, low pressure or deep reservoirs. In fact it is the only thermal method that can presently be applied to deep reservoirs, though deep down steam generation is being tested. It can be used at any stage of reservoir depletion. It can be used in special situations such as offshore or in arctic regions. Because of the lack of heat losses at the surface and in the injection wells, it is the most thermally efficient thermal recovery method. The inject ant (air) is readily available. Combustion allows wider well spacing than steam. Economic results are comparable to those of steam injection.

Several aspects of operating ISC projects are important. First is the large compression ratios and associated costs required to inject air into the formation. Second is planning and design requirements for a combustion project that are more difficult than steam. Third is extensive laboratory work to assess fuel availability, air requirements and burning characteristics of the crude that are required before designing in situ combustion projects. Fourth is the high degree of technical sophistication and careful monitoring needed to ensure proper operation of project Fifth is the limitation of numerical simulation and other techniques that make predictions of recovery more difficult than most other enhanced oil recovery methods .As mentioned before, considerable improvements in the application of in situ combustion have been made since the early projects .New developments like using oxygen enriched air, injecting water containing surfactant and alternate slugs of air during forward combustion etc improve the efficiency of fire flooding as well as eliminate the problems faced during its application.

352165 Design of Surfactant-Only Chemical Flooding in a Low Perm, High Tds Reservoir: Single Well Test

Abstract:

Low permeable oil reservoirs are often characterized by slow waterflood injectivity and low fluid productivity. For low perm formations, the permeability appears to be the limiting factor to deploy Surfactant-Polymer (SP) and Alkaline Surfactant Polymer (ASP) flooding for improved oil recovery. Injection of carbon dioxide has shown promising results in low perm formations and is commonly used in these fields when sufficient CO_2 can be obtained. However, even when sufficient CO_2 can be sourced, initial capital investments for CO_2 injection may be quite high and may limit its application for some mature fields. In this paper we report a single-well test of an improved surfactant-only formulation in a low permeability, high TDS reservoir. The target single well, SEH1, is located in south central of Oklahoma. The perforated zone is located 6,000 feet below the surface with an on-going waterflood. The average permeability of the formation is 14 mD, ranging from 0.2 to 44 mD. The produced brine contains total dissolved solids (TDS) at 102,300 mg/L and the reservoir temperature is 50 °C (123 °F). The reservoir crude exhibits a viscosity of 23 cp at reservoir conditions, presenting further challenges for an EOR process without mobility control using polymer. For the test, the hydrophilic-lipophilic difference (HLD) equation was used to predict possible salt-tolerant surfactant formulations, based on site-specific fluid properties and reservoir conditions. Among the binary and ternary surfactant mixtures developed, the surfactant formulations were further optimized by introducing low concentrations (between 400 to 600 mg/L) of hydrophilic linkers to achieve a stable surfactant solution in the reservoir brine and ultra-low interfacial tensions (< 0.008 mN/m) between site crude and produced brine. In laboratory testing injection of three pore volumes of a surfactant-only formulation in sand packs and core flood devices achieved tertiary oil recovery ranging from 45% to 70% of the residual oil (Sor) after water flooding. The levels of surfactant loading were varied from 0.23

wt% to 0.92 wt% to assess the impact of surfactant concentration along with slug size. The field single well test used a ternary surfactant formulation (i.e., a mixture of alkoxylated alcohol sulfates/ethoxylated alcohol sulfates/blends of monoalkyl and dialkyl diphenyloxide disulfonates) and was conducted in April-May of 2013 to further confirm laboratory results at the SEH1 site. The 3-week single-well test was a technical success. A surfactant only formulation was injected at 0.69% followed by injection of reservoir brine without mobility control. Based on the pre- and post-chemical tracer test results (i.e., use of ethyl formate as partitioning tracer plus other conservative tracers), approximately 90% of the residual oil at the target zone was mobilized. This field experience will be built into a design protocol using single-well tests to mitigate the risks of pilot-scale chemical flooding targeted at mature fields in the U.S. mid-continent region.

352753 Foam Mobility Control: An Enhanced Oil Recovery Technique

Abstract:

In Enhanced Oil recovery techniques a significant fraction of the oil-in-place is left in the ground after primary or secondary recovery due to various reasons. Gas injection is one method used to extract these oils. There are basically four types of gases used in gas injection: natural hydrocarbon gas, inert gas, carbon dioxide, and steam; but in all cases there is the problem of gravity override.

Gravity override meaning the preferential movement of the injected gas through the upper part of the reservoir due to the lighter density of gas compared with oil. Gravity override causes early gas breakthrough in production wells and hence, reduced oil recovery. The steam is injected into the ground to reduce the viscosity of the oil and to make the oil more mobile. Unfortunately, due to gravity over ride, as steam rises to the top of the reservoir. Together with reservoir heterogeneity, gravity override results in early steam production and reduced volumetric sweep efficiency. Hence recovery is not improved to the degree desired. The injection of a permeability blocking agent would do the required. Various scientist have reviewed the selective blockage of fluids in thermal recovery projects and concluded that foam is best suited for this purpose. The foam is found to be promising to work in alkaline/surfactant processes to replace or supplement polymer as mobility control agent. The present study deals with the advancements in the foam mobility control for the recovery of unrecovered crude oil from the reserve.

352632 A Qualitative in Field Test to Characterize Oil-Shale Reservoirs

Abstract:

Shale oil recovery, despite its extraction costs and low yield per shale matrix is becoming a more feasible solution to the rapidly emerging global fuel demands. However, many oil-shale reserves are located several thousand feet below the earth's surface, which makes it extremely difficult and expensive to extract. In addition, the quality of the shale matrix may have a broad range such that the potential yield through and other methods may vary significantly across neighboring reservoirs. And determining the quality of the oil-shale matrix and its potential yield of synthetic/tight oil using laboratory analytical methods can take several weeks. This work presents a rapid and simple qualitative "in-field" test-tube test method that categorizes and prioritizes a given shale matrix based upon its composite. This test method can be a significant cost benefit and a proof positive indicator of where to focus exploration and recovery interest.

Pressure Regulation and Kick Attenuation with Wired Drill Pipe Technology in Managed Pressure Drilling

Hector Perez
Reza Asgharzadeh Shishavan
John Hedengren
Brigham Young University

Outline

- Overview
- Motivation for Pressure Control
- Method and Structure of the Control System
- Results
- Results of Automated BHA Pressure

Motivation for Pressure Control

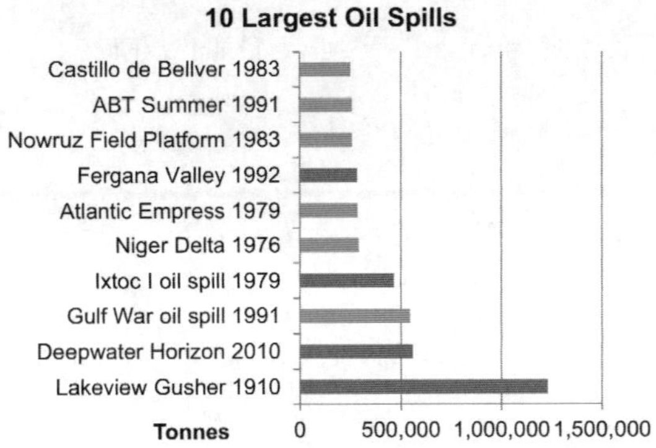

Apr 2, 2014 Pressure Regulation and Kick Attenuation with Wired Drillpipe in MPD

Motivation for Drilling Control

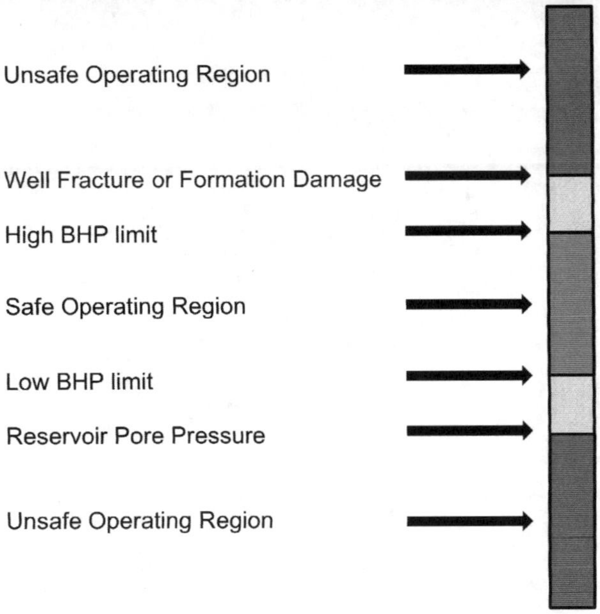

BHP Control Objectives

Pressure Regulation

Kick Attenuation

Apr 2, 2014 Pressure Regulation and Kick Attenuation with Wired Drillpipe in MPD

Managed Pressure Drilling System

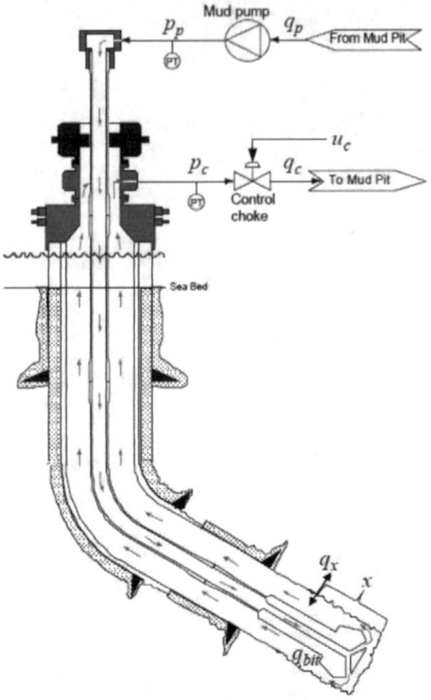

- ❑ Overbalanced Drilling
- ❑ Managed Pressure Drilling (MPD)
- ❑ Underbalanced Drilling

MPD offers:

- ❑ More precisely control of borehole pressure
- ❑ Decrease NPD and mitigate drilling hazards

Manual Control versus Automated Control

Challenges	Benefits of BHP control Automation
Large Time Delays	Improve Reliability of Pressure control
Complex Behavior of Multiphase flow	Reduce the possibility of human errors
Sudden Change in Formation Properties	

Objective

Designing a comprehensive controller which Simultaneously

- Regulates the BHA Pressure within specified range
- Attenuates Unwanted gas influx when it happens
- Uses latest automation technologies
- Has improved performance in comparison with manual and conventional controllers

Wired Pipe Technology to be used in Control Scheme

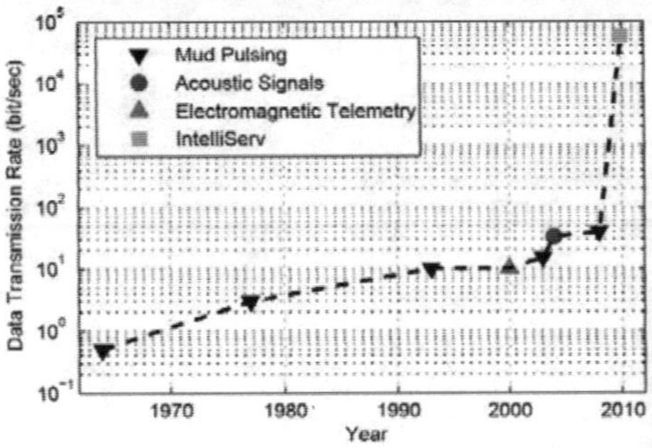

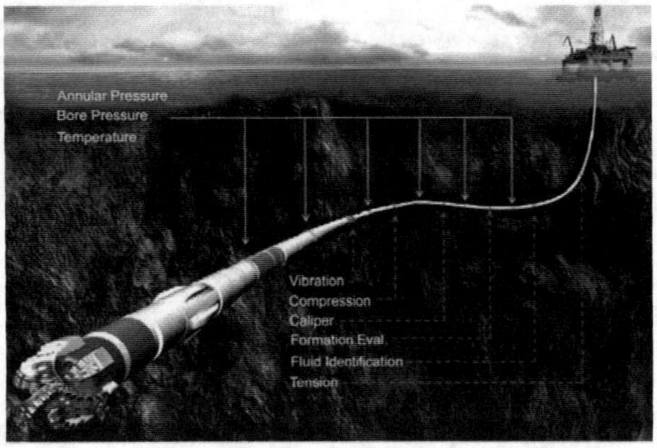

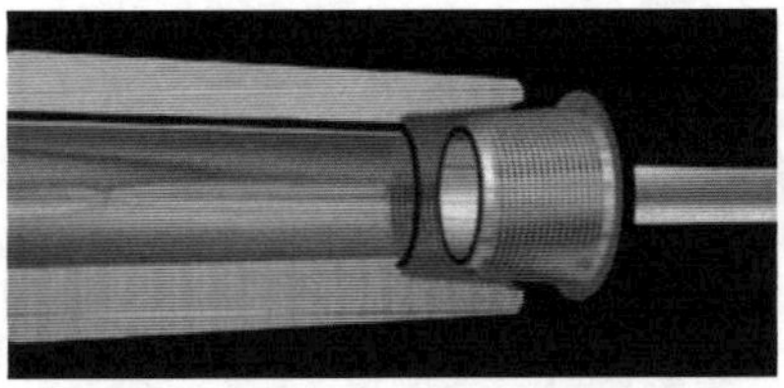

Challenges in the designed controllers

Mud Pulse Telemetry Based	Wired Pipe Telemetry Based
❑ Indirect control of BHP	❑ Directly control of BHP and annular pressures ✔
❑ High uncertainty in estimated variables values	❑ Increased precision in estimating the unmeasured process variables ✔
❑ Still Large time delay in kick detection	❑ Improved kick detection ✔

Schematic of the MPD Controller

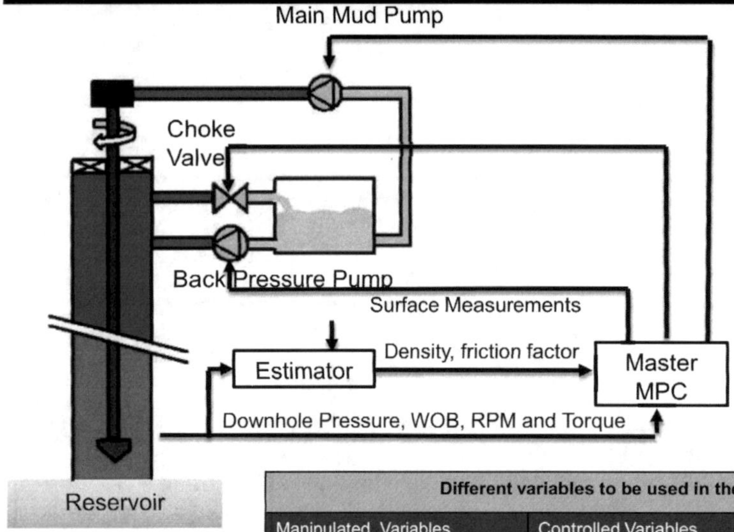

Different variables to be used in the current research		
Manipulated Variables	Controlled Variables	Estimated Variables
Rig Pump Flow Rate	BHP	Annulus Drill Fluid Density
Choke Valve Opening	Annular Pressures	Annulus Friction Factor
		Reservoir Pore Pressure

Apr 2, 2014 Pressure Regulation and Kick Attenuation with Wired Drillpipe in MPD

Controller Structure

Multivariate Nonlinear Model Predictive Controller (MPC)

What are advantages over conventional controllers?

Nonlinear Feature of the Controller

Multivariate Nonlinear Model Predictive Controller (MPC)

- Drilling is a non linear process
- Current controllers use linear (simplified) models
- A nonlinear controller can capture the dynamics better
- Designing nonlinear controllers are more challenging

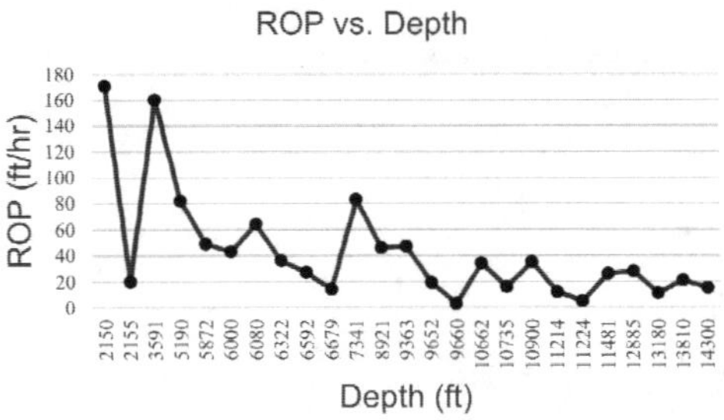

Figure 1. Change of ROP with drilling depth. Field data from (Kinabalu East-1, 1990)

Model Predictive Control and Moving Horizon Estimator

Controller

Multivariate Nonlinear Model Predictive Controller (MPC)

- Future prediction capability – Improved control performance
- MPC vs. Conventional Controllers

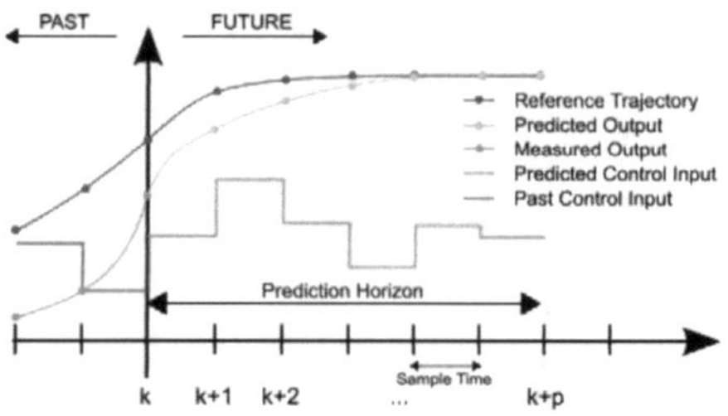

Estimator

- Moving Horizon Estimator
- Uses both downhole and surface measurements
- More precise and faster parameter and state estimation

Software

- APMonitor Optimization Suite
- Designed Especially for Dynamic Nonlinear Processes
- Using the Most Efficient Solvers Available
- Solution Time: Fraction of a Second

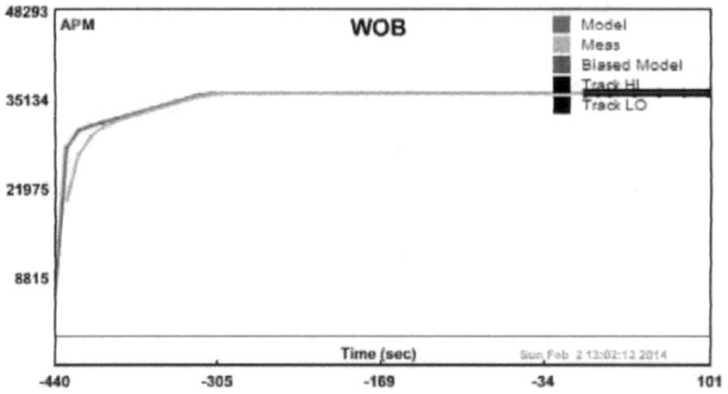

Lower Order Model

Balance Equations

$$\dot{p}_p = \frac{\beta_d}{V_d}(q_{pump} - q_{bit})$$

$$\dot{p}_c = \frac{\beta_a}{V_a}(q_{bit} + q_{back} - q_{choke} + q_{res})$$

$$\dot{q}_{bit} = \frac{1}{M}(p_p - F_d|q_{bit}|q_{bit} + \rho_d g h_{bit} - p_{bit})$$

$$\dot{\mathbf{h}}_{bit} = \boldsymbol{ROP}$$

Closure Relations

$$M_a = \rho_a \int_0^{l_w} \frac{1}{A_a(x)} dx$$

$$M_d = \rho_d \int_0^{L_{dN}} \frac{1}{A_d(x)} dx$$

$$M = M_a + M_d$$

$$p_{bit} = p_c + \rho_a \times F_a|q_{bit} + q_{res}|(q_{bit} + q_{res})h + \rho_a g(h_{bit} - (n-1)h)$$

$$q_{choke} = K_c z_{choke}\sqrt{\rho_{an}(p_c - p_0)}$$

Zhou, Kaasa, et al. (2008

Kick Detection and Attenuation

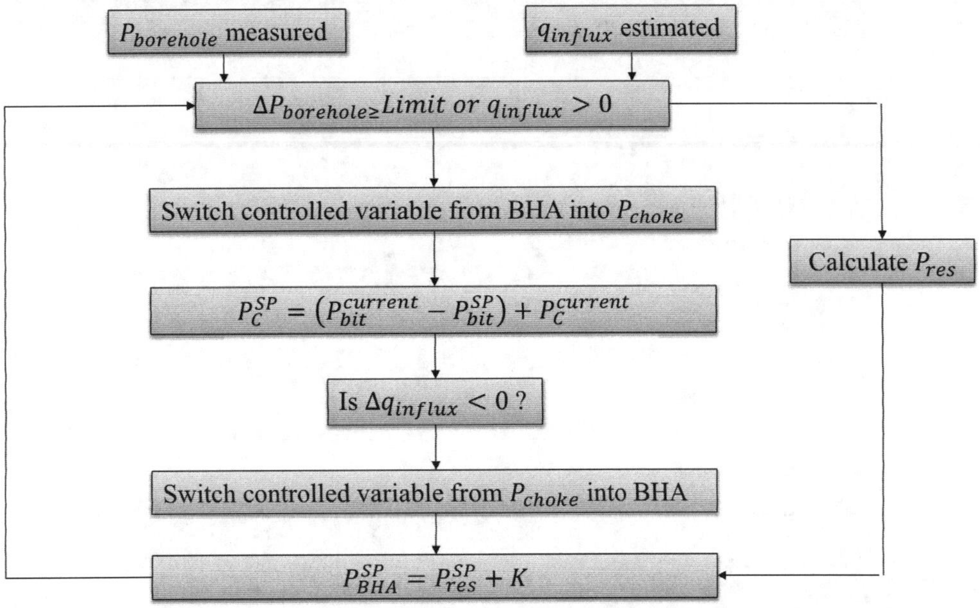

Results

Automated BHA pressure control

Table 1. Initial well conditions used for simulations

Variable	Value (SI units)	Value (US units)
Distance between two successive pressure sensors h_a	150 m	492.1 ft
ROP	0.01 m/s	0.033 ft/s
Fluid density in the drill string - ρ_d	0.012×10^{-5} kg/m^3	74.9 lb/ft^3
Fluid density in the annulus - ρ_a	$0.014^a \times 10^{-5}$ kg/m^3	87.4 lb/ft^3
Bulk modulus in the annulus β_d	14000 bar	2.03×10^5 psi
Bulk modulus in the annulus β_a	14000 bar	2.03×10^5 psi
Annulus cross sectional area	0.021 m^2	0.226 ft^2
Friction factor in the drill string F_d	10000 bar m^{-6} s^2	116.3 psi ft^{-6} s^2
Friction factor in the annulus F_a	1250^a m^{-5}	3.29 ft^{-5}
Lumped density per length parameter for the drill string - M_d	5700×10^{-5} kg/m^4	108.5×10^{-5} lb/ft^4
Lumped density per length parameter for the annulus – M_a	800×10^{-5} kg/m^4	15.22×10^{-5} lb/ft^4
Well depth (h_{bit})	2000 m	6562 ft
[a] For connection procedures, the values used for simulation were 0.01 (density) and 1000 (friction factor)		

MHE Estimator

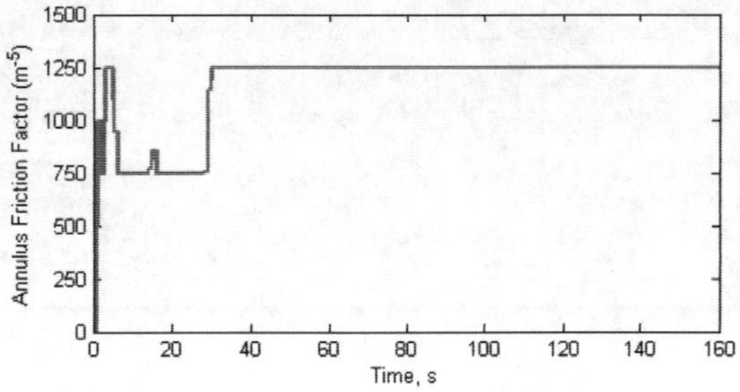

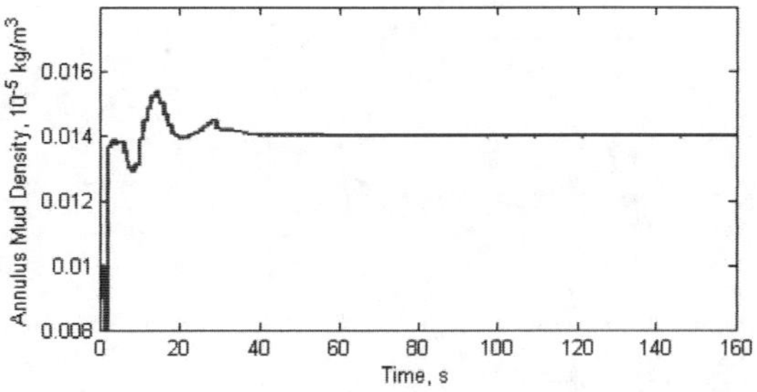

Within 30 seconds of operation, the estimator has accurately determined the friction factor and has determined the annulus density within 1.5% of the actual value.

Case 1: Normal Operation

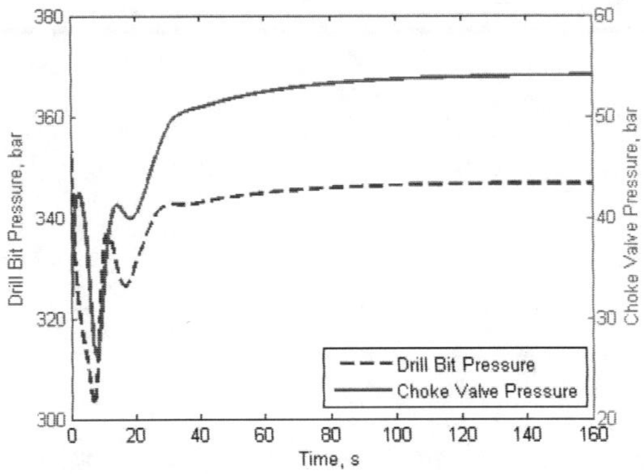

Within 30 seconds, the drill bit pressure has been controlled within 0.72% of its set point of 345 bar

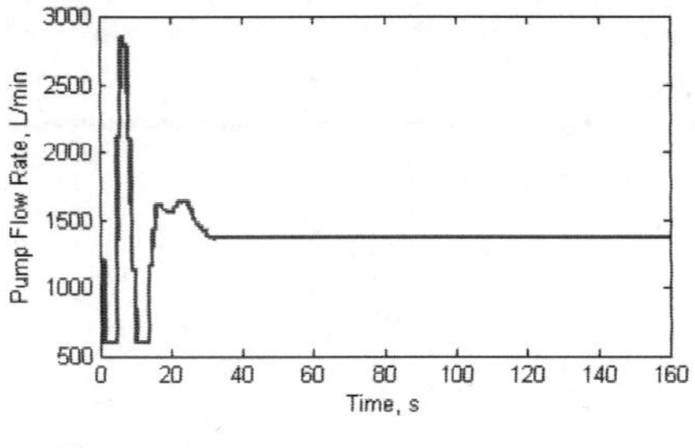

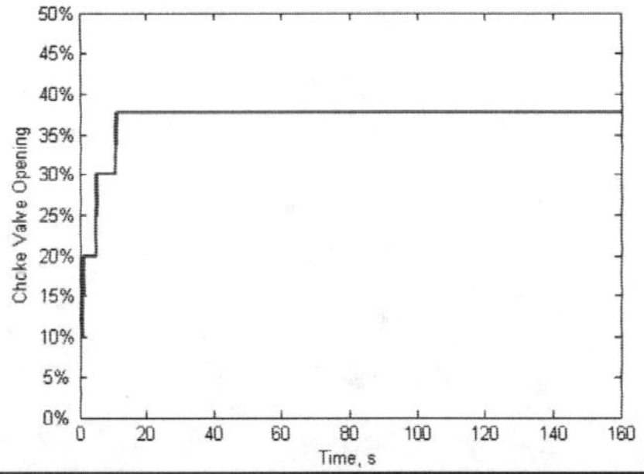

Case 2: Connection Procedure

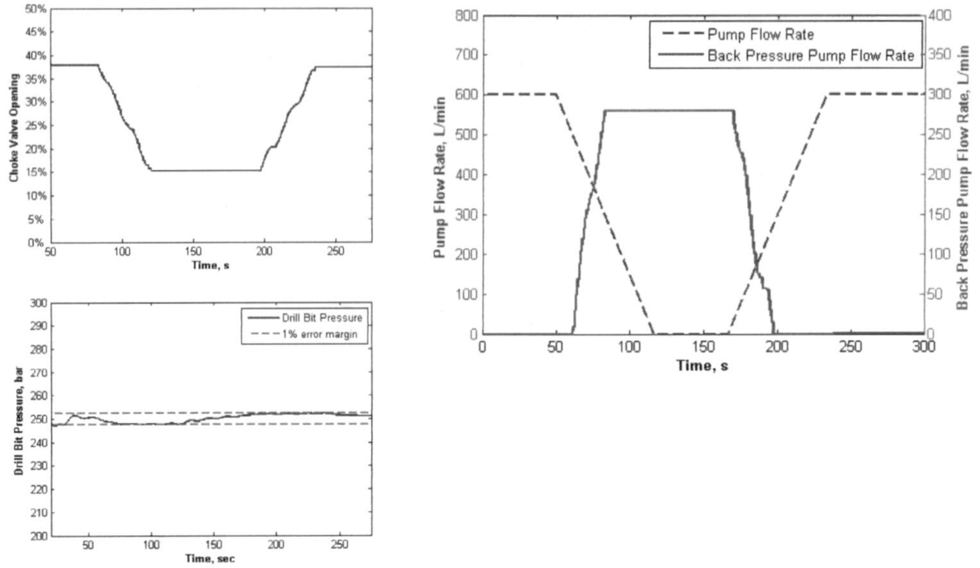

Case 3: Unwanted Gas Influx

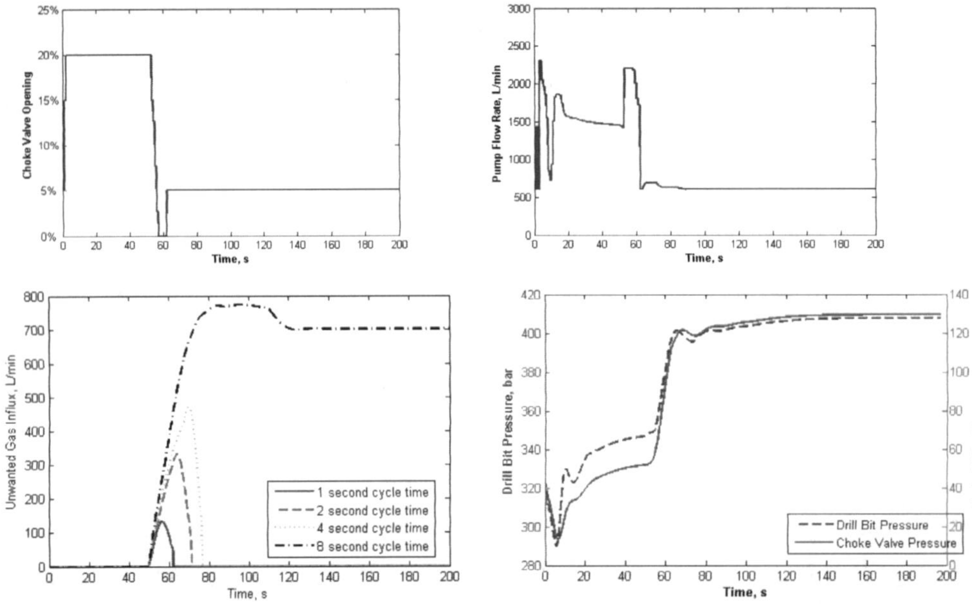

Thank You for Your Attention

Apr 2, 2014 | Pressure Regulation and Kick Attenuation with Wired Drillpipe in MPD

Spring Meeting, New Orleans, LA

Risks from Unconventional Wells Gasification

Dennis B. Butts
Americas Risk Practice
Environmental Resources Management (ERM)
CityCentre Four
840 West Sam Houston Parkway North, Suite 600, Houston, TX 77024
dennis.butts@erm.com

Prepared for Presentation at
American Institute of Chemical Engineers
2014 Spring Meeting
Topical 2: Shale Gas and Tight Oil
New Orleans, LA
March 30 – April 2, 2014

UNPUBLISHED

Risks from Unconventional Wells Gasification

Dennis B. Butts
Americas Risk Practice
Environmental Resources Management (ERM)
CityCentre Four
840 West Sam Houston Parkway North, Suite 600, Houston, TX 77024
dennis.butts@erm.com

Keywords: LNG, Risk Management, Process Safety, Unconventional Wells, Shale

Abstract

With the continued differential between diesel and natural gas costs, onshore well operators have been working to convert both drilling rigs and hydraulic fracturing pumps to utilize natural gas as fuel. Associated with this conversion is a change in the risk profile that may not be fully recognized or understood in the field. Risk management philosophies and assumptions need to adjust to the difference in hazard characteristics between the fuels, as well as the variety of delivery methods.

Specific topics will include:

- Motivations for fuel conversion and the differences in scale for rig power, natural gas substitution on diesel hydraulic fracturing pumps, and dedicated natural gas powered hydraulic fracturing pumps. Risk mitigation and field training requirements may not fully accounted for in investment decisions.
- Scale of conversion and location specific availability drives natural gas sourcing, which can then drive personnel risk and well pad layout issues depending on whether the gas is sourced from local wellheads, pipeline, Compressed Natural Gas (CNG), or Liquefied Natural Gas (LNG). Separate Simultaneous Operations (SIMOPS) issues can arise based on the specifics of each sourcing method.
- Associated risks need to be appropriately managed in a manner that involves the personnel directly exposed. And there is a need to ensure that well sites which did not previously present a threat to surrounding activities do not become a threat due to either direct risk from the fuel conversion itself or from escalation risks not previously of concern.

1. Introduction

Over the past few years there has been a significant increase in efforts to substitute natural gas for diesel to provide fuel for drilling and fracturing onshore wells. This is an expansion of the conversion efforts that were already underway to convert transportation fleets away from diesel fuel. While the reasons for this may vary, and do include environmental benefits, the most significant reason is cost savings.

Natural gas has been averaging about $4.50/mmbtu [1] over the past 6 months, which is about $0.59 per diesel gallon equivalent (DGE) [1]. Even with added distribution costs, if pipeline gas is available, the economics are very attractive. Wellhead gas is even more attractive. And if LNG is the only alternative at a well site, at $4.50/mmbtu, the natural gas component of LNG is about $0.34 per gallon [1], with a markup of about $0.15 per gallon for liquefaction and storage [3] plus transport costs of about $0.15 per gallon (up to 100 miles) [4], giving a breakeven price of $0.64 per gallon or $1.10 per DGE. With typical supplier pricing to bulk customers as a discount to diesel typically around $1.50 per gallon, and more substantial discounts for large scale contracts, there are significant financial benefits for both LNG suppliers and drillers when diesel fuel is in the $4 per gallon range.

And as a bonus to the economic justification, drillers can tout emissions reductions of as much as 20 – 30% relative to diesel [5]. Rarely in these discussions is the subject of added costs or effort for process and technical safety brought up.

Substitution of natural gas for diesel has been underway for some time in heavy vehicle transportation fleets. With added competition in engine conversion markets and among storage tank suppliers, conversion costs have been reduced to the point where heavy vehicles can recover costs in less than 2 years [6]. Efforts to convert drill rigs originated in the Marcellus region, due to environmental concerns, but the economics have driven natural gas rigs into most of the major markets in the US, as well as into western Canada. Since hydraulic fracturing pumps use even more diesel than rigs, recent efforts have been in this direction, though technical challenges regarding power delivery have resulted in two types of conversion. One is the substitution of gas for diesel in existing engine drivers to ensure sufficient horsepower. The other is to use high powered turbine pumps, which are very well suited to gas power. Large scale conversion of fuel sources is not unprecedented, as the US heavy truck fleet went from less than 10% diesel powered to nearly 100% diesel powered from 1950 to 1980 [6].

So what does this mean in terms of risk management? Aside from the business and financial risk of converting equipment to a fuel in natural gas that is historically more volatile in pricing than diesel fuel at a time of historically high oil to gas price differentials, which is a subject for another paper, there are practical technical safety considerations that come into play as drilling sites convert from diesel to natural gas. These can result in increases to the risk profile of a well pad that is not generally accounted for, as well as changes in the risk profile within the drilling operation itself.

This paper will discuss some of the issues and options regarding conversion and highlight specific risk issues associated with each.

2. Sources of Natural Gas

2.1 Typical Volume Requirements

A heavy truck may use 20,000 gallons of diesel per year [4], with significant potential savings being the result of a conversion to natural gas fuel. Conventional fluid drill rigs use an average of 425,000 gallons of diesel per year [5], creating a fuel cost savings equivalent to converting a fleet of over 20 heavy trucks, but at an investment cost that is much lower. But even these volumes pale in comparison to the amounts of diesel used in hydraulic fracturing. Fracking crews can use 5,000 to 10,000 gallons of diesel per day [7], and it is common for the crews to be in operation for as many as 300 days per year, accounting for mobilization and demobilization time. That means 1.5 to 3 million gallons of diesel per year can easily be used by a single crew, often going well to well for the same operator. An unfortunate outcome of this is that conversions become a foregone conclusion, without due consideration to the resulting process safety risks.

2.2 Options for Sourcing and Delivery

Natural gas for a well site can be sourced as gas directly from the wellhead or gathering system, or it can be piped in from a commercial distribution line, when those are available. In more remote locations, distribution lines may not be available, and if the gas is too wet, wellhead or gathering gas may also be unsuitable for use. In those situations the options are to utilize either CNG or LNG. CNG would require considerably more transportation than LNG due to the lesser amount of gas that can be moved and comes with higher costs for storage and handling equipment. Therefore, CNG has limited utility for well sites, though its lower fuel cost makes it useful in some narrow applications. The use of LNG can be cost effective when other options are not available, as it is more available than in the past and although containing less energy than diesel per gallon, an onsite tank containing 15,000 gallons of LNG is still equivalent to over 8600 gallons of diesel [2] and is about 2.4 times the energy density of CNG [9].

2.3 Issues Associated with Each Option

Use of wellhead gas or gas from gathering lines requires either very dry gas or extensive treatment, as most engines converted to utilize natural gas are designed for gas approximating domestic gas. This is in contrast to offshore facilities where engines are often optimized to take advantage of fuel gas diverted from separation trains. Pipeline gas can be the best option, for those sites with ready access to a distribution line, and will provide sufficient flow and pressure for most natural gas drilling rigs. However, pipeline gas may have to have compression added for certain types of fracking pumps, and available flow may not be sufficient for some fracking operations. CNG can be useful in areas that do not have dry gas or ready pipeline access and may not have easy access to LNG. In such areas, CNG facilities can be set up to provide the product to users in the surrounding region much more readily than LNG facilities. To utilize CNG does require more transportation of the gas and more onsite storage than using LNG. With LNG being the last supply option, but becoming very common in the field, this is an option that carries significant process and technical safety risks. LNG is a material than can bring many benefits, but needs to be treated with considerable respect, often more than is witnessed at well sites.

3. Associated Risks

Associated process and technical safety risks arise in several stages of the conversion from diesel to natural gas. In most cases there is an increase in those risk levels relative to diesel fuel, though this is not always fully contemplated by decision makers or well communicated to the field level staff. Whereas diesel fuel is classified as a combustible liquid, NFPA Class II [8], is difficult to ignite if spilled and primarily a localized threat, natural gas is highly flammable vapor and depending on delivery method could have the potential to disperse considerable distances if released.

In use in equipment, natural gas can create the need for hazardous area class or hot work permit zones that were not required around diesel equipment. This issue has been missed on well sites and often is not included in discussions about conversion costs until after being identified in PHA studies or other risk assessments. The same consideration extends to gas monitoring and shutdown systems. System risks are manageable, but need to be given due consideration and treated with the proper respect. Some expansion of this thought is provided below.

3.1 Source of Gas

Wellhead gas primarily introduces risk through the separation and dehydration process. Even relatively dry gas needs to be treated prior to use in most of the engine drivers used for onshore production. To utilize it during drilling usually means that at least one well has already been put into operation on a multi-well pad. The risks associated with production while drilling and completing adjacent wells have to be assessed. Controls that may need to be in place during certain phases of drilling could interfere with the supply of gas. An alternative that may be available within a field of well sites under one operator is to use gas from the gathering system. While treatment would also be required for that gas, the potential SIMOPS issues of a producing well on a pad with drilling activity. In either case the gas being treated will be a relatively high pressure and is likely to result in some amount of condensate or natural gas liquids (NGL) production and handling. This can increase the fire and explosion risks above those normally found during drilling activity.

Pipeline gas is only an option when a distribution line is nearby. But if available, this can provide a supply of natural gas that is already treated and can be adjusted to the necessary pressure at the takeoff location from the distribution network. In analysis of fire and explosion risks, pipeline gas supply most often has the smallest consequence zones and the least amount of equipment susceptible to failures. This is even the case if high horsepower fracking pumps are in use that requires some degree of compression of the natural gas to reach necessary operating pressures.

CNG introduces risk due to its high pressure, typically near or above 3000 psig [9]. This can both drive the spread of a gas cloud in the event of a release as well as provide the means through depressurization to cool the gas to a low enough temperature that it will disperse for some distance along the ground, further lengthening dispersion distances. Storage under pressure also introduces a finite risk of boiling liquid expanding vapor explosion (BLEVE) that could spread debris.

LNG usage at well sites is capable of imposing a larger potential impact zone than a well blowout scenario. And this is excluding a catastrophic failure of the storage medium, which is low likelihood event due to the dual tank construction of trailers that handle LNG. When released, LNG will initially disperse in a manner very similar to liquefied petroleum gas (LPG), but as the edges of the dispersing cloud warm, flammable vapors will loft upward. Assumptions regarding layout and spacing that are based on LPG can address the flammable breadth of a possible LNG cloud, an example of which is shown in Figure 1 from PHAST [10], going out about 850 feet (260 meters) to the 50% lower flammable limit (LFL). Maximum distance to a 1 psi overpressure, good for window breakage, trailer damage, and upsetting neighbors, would extend another 300+ feet (100 meters) beyond the 50% LFL distance. In general, the height of the 50% LFL section of the cloud will remain under 15 feet (5 meters), but this also needs to be confirmed relative to the drill rig flare height. And the failure of a LNG loading hose would impose an even larger impact zone. Though many operators end up using gas detection, this is only effective if appropriate emergency response procedures are in place.

Figure 1: 50% LFL Dispersion Distance for 1-inch LNG Leak [10]

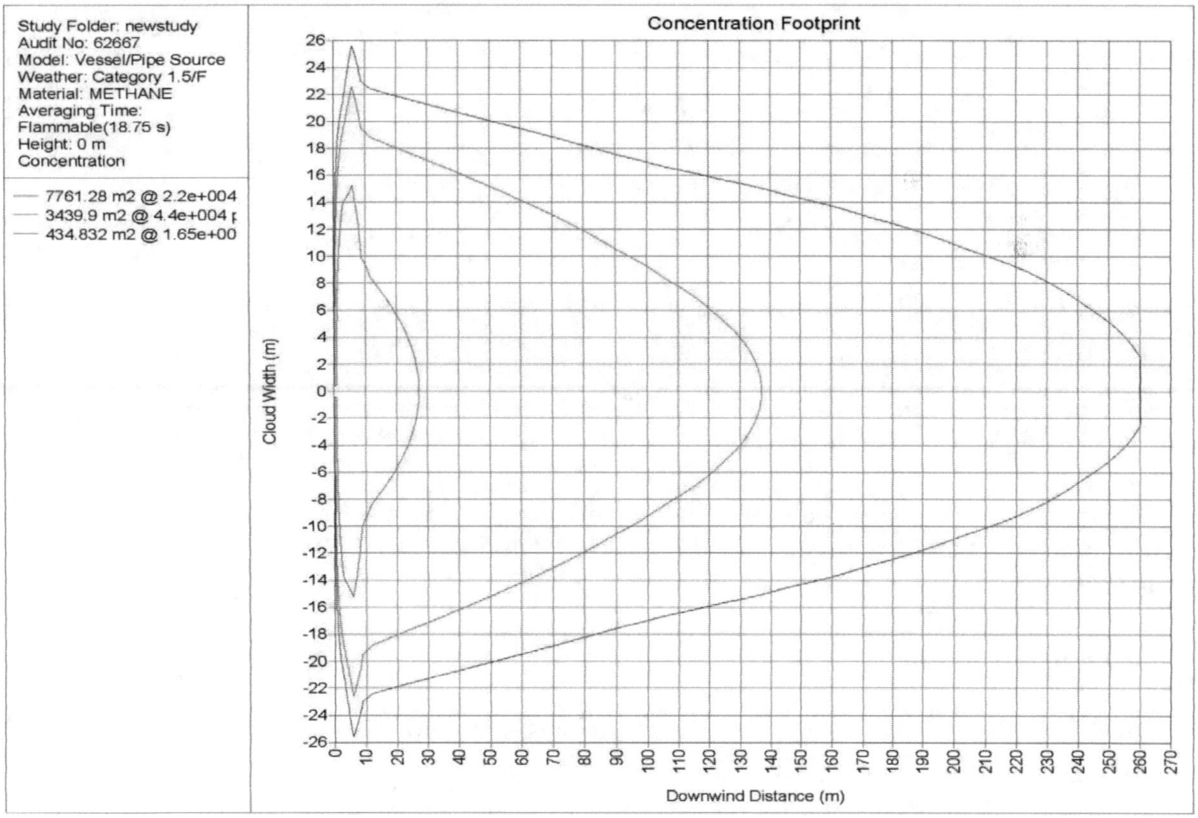

For LNG most releases behave as impinged releases, due to the low pressure environment. For CNG, the best case is for a free field release, which provide dispersion distances of less than half that of LNG. However, horizontal releases of CNG are capable of dispersion distances beyond those of LNG, in particular if they impact other equipment and become partially impinged, as shown for dispersion to 50% LFL for a 1-inch CNG release in **Figure 2**. An interesting phenomenon is how the LNG dispersion slumps outward from the beginning, due to release at temperatures low enough for the methane to be denser than air, but the CNG is initially released at warmer temperatures, then settles into a much lower temperature, resulting in a slumping spread.

Figure 2: 50% LFL Dispersion Distance for 1-inch CNG Horizontal Leak [10]

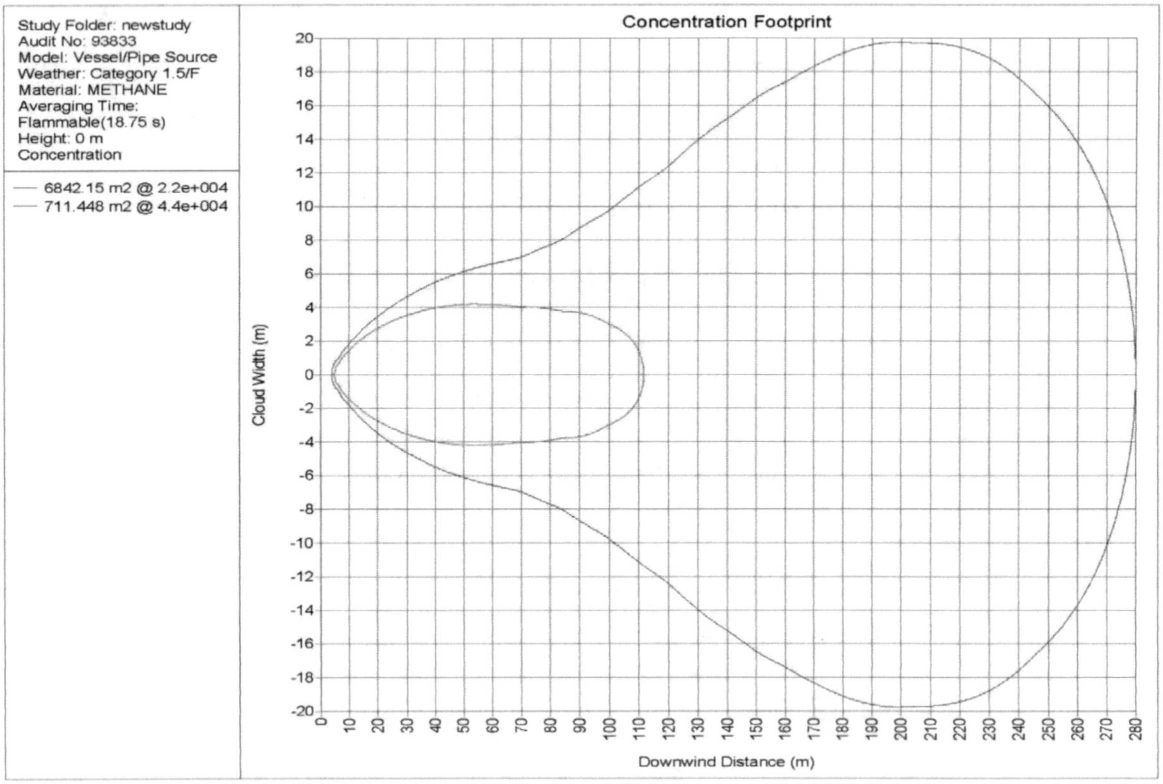

As a reference, an ignited well blowout, traditionally considered a worst case event at a well pad, will typically have an effect distance to 1500 Btu/hr-ft^2 (4.7 kW/m^2) of about 500 feet (150 meters).

3.2 Refueling at Site

When utilizing CNG or LNG as the source of gas, there are additional considerations for storage at the site and for refueling deliveries. Both require delivery by truck, just like diesel fuel, but the release of diesel from an overturned truck is not the same as a release of compressed or liquefied natural gas. This is why there are stringent design requirements on CNG and LNG transport vehicles. However, when assessing risks it is wise to keep in mind that those standards were developed for paved highway transportation, not gravel access roads to well sites. In addition to possible visibility issues, there are the impacts such roads could have on the expected behavior of safety features on the vehicles.

3.3 Compression and Delivery of Gas to Equipment

Further issues arise in situations where either pipeline gas or LNG is used to supply equipment such as hydraulic fracturing pumps that require higher pressure feed gas. In these circumstances, arrangements have to be made to provide compression at the site. A site that is typically already very crowded with equipment. In terms of risk from the equipment it is normally a localized threat, however the extra equipment adds a level of complexity that can complicate operations. The added complexity can add a human factors risk dimension that is rarely assessed in drilling situations.

4. Conclusion

Considerable financial and environmental benefits can be had by converting drilling operations from diesel fuel to natural gas, as long as the current differential in crude oil versus natural gas pricing exists. To achieve those benefits does entail some risk, with this paper having focused on the process and technical safety risks. In some instances, the benefits and risks can pull in opposing directions creating a challenge to find equilibrium between the two.

Depending on the source of natural gas for the drilling and completions operations, due consideration needs to be given to:

- Training needs for field staff that are used to dealing with combustible diesel fuel and not highly flammable natural gas;
- Flammable gas dispersion distances as well as resulting maximum potential zones for 1 psi overpressures, in particular for LNG or CNG applications;
- Review of drill rig flare location relative to natural gas supply point(s) to account for potential height of dispersion;
- Road conditions and their potential impact on transportation of fuels;
- Hazardous area classification zones and resulting equipment alterations, layout changes, or work permit requirements;
- Gas detection needs and the emergency response planning to accompany the detection; and
- Hazard assessment requirements and when they are activated to address changes made to drilling operations or well pad layouts once in service.

As we further explore this question, it is necessary to keep in mind not only the benefits of the ongoing process of converting away from diesel fuel to natural gas at well sites, but also the risk management challenges that can arise.

5. References

[1] US Energy Information Administration, 2014

[2] US Department of Energy, Fuel Energy Spreadsheet, 2014

[3] Quine, Tom, "Fundamentals in LNG", ASGMT White Papers, 2010

[4] Center for Energy Economics, University of Texas at Austin

[5] Oil & Gas Technology, "Gas Powers Fracking Plays", April 10, 2013

[6] Blomerus, Paul and Oulette, Patric, "LNG As a Fuel for Demanding High Horsepower Engine Applications: Technology and Approaches",

[7] Schutt, Bryan, "Diesel to Gas Switching Trend Takes Off in Drilling Sector", SNL Financial, September 3, 2013

[8] NFPA 30, National Fire Protection Association, 2012

[9] Envocare Ltd., "Liquefied Petroleum Gas (LPG), Liquefied Natural Gas (LNG), and Compressed Natural Gas (CNG)", 2007

[10] PHAST 6.7, DNV GL AS

353108 Supercritical CO_2 / Brine / Hydrocarbon Interactions in Shale Rock Fracture: A Geo-Microfluidics Approach

Abstract:

Shale oil and gas is a critical unconventional resource for the energy sector in the US and worldwide. The booming of its production in the recent decade, due to the hydraulic fracturing technique, has led to serious environmental concerns such as huge water demands and toxic additives into fracking water which may pollute groundwater. A proposed solution is to use supercritical CO_2 ($scCO_2$) as a more environment-friendly alternative fracking fluid to replace water for enhanced oil recovery (EOR) in oil shale. Many challenges lie in the way of this scheme such as to understand multiphase flow of $scCO_2$ at injection sites, its interactions with brines, hydrocarbons, and surrounding geologic matrix. We present here a microfluidic approach to investigate the micro-scale multiphase fluid flow and reactive transport within low-permeability shale rock. Unlike the conventional microfluidic techniques, the geo-microfluidic chip is directly fabricated from a thin-section of shale rock, hence provides the genuine surface properties. Simple fracture-like patterns are created on the shale rock surface by machining and femto-second laser etching to simulate the natural structures. Flow experiments in the geo-microfluidic chip are conducted within a high-pressure, high-temperature apparatus to simulate the in situ reservoir conditions. The interactions between $scCO_2$/brine/hydrocarbon will be quantified and the hydrocarbon extraction efficiency of using $scCO_2$ and using fracking water will be compared. The displacement patterns, the dissolution rate of $scCO_2$ in brine/hydrocarbon, and the surface wettability of $scCO_2$ and shale rock will also be investigated. These micro-scale experiments results will help to reveal the underlying mechanisms of multi-phase fluid flow under in situ temperatures and pressures.

352468 Model Shale Gas Properties from a Statistical Mechanical Perturbed Hard-Sphere Chain Cubic Equation of State

Abstract:

Shale gas is an increasingly important energy source in North America. Hence, an accurate description of the physical, transport and thermodynamic properties of shale gas mixtures are a key prerequisite for process design. A particularly advantageous feature is that multiple properties can be obtained from a single equation of state (EOS) for a model system. Traditional EOS's are cubic in molar volume and are derivatives of the basic van der Waals (1873) EOS. Redlich and Kwong (1949), Soave (1973) and Peng and Robinson (1976) are attempts to improve the accuracy of the basic vdW EOS; however, as is shown early in this talk, these equations of state overcorrect the attractive term to compensate for deficiencies in the repulsive term. On the other hand these EOS's have the advantages of a computationally simple form and *three well-defined molar volume roots*. Nonetheless, while these EOS's often give a reasonably accurate representation of the phase composition of simple binary and ternary alkane mixtures (model shale gas systems); poor results are obtained for other properties including liquid density, mixture second virial coefficient and diffusion coefficient.

On the other hand, a class of accurate molecularly-based but complex and computationally intense equations of state: the statistical associating fluid theory (SAFT) equations (Huang and Radosz, 1990; Chapman et. al., 1990), which account for individual molecular level effects (i.e., hard-sphere repulsion, London dispersion, covalent bond chain formation, multipolar forces and hydrogen bonding) have successfully emerged in the past few decades to account for multiple effects in pure fluids and mixtures including associating and/or polydisperse components. These EOS's capture the major molecular level effects; however, convergence difficulties and the existence of several molar volume roots complicate implementation of these EOS routines.

Consequently, considering the above difficulties, we have developed an approach which incorporates the rich molecular description of the SAFT equations into a cubic in molar volume form. Specifically, molecular simulation data for hard-sphere compressibility factor is nearly exactly fit to a simple empirical function in packing fraction, which ensures an overall cubic in molar volume equation of state and from which two universal parameters (applicable to all fluids and fluid mixtures) are obtained. Furthermore, London dispersion attractive forces are described by an extremely simple polynomial refit of data generated for Chen and Kreglewski's (1977) *thirty-six term* double sum in reciprocal absolute temperature and packing fraction. Finally, chain formation (covalent bonding) is described by a modified form of Wertheim's (1984) thermodynamic first-order perturbation theory (TPT1) for segment covalent bonding, in which a diimer reference radial distribution function is employed to obtain more accurate results. Conventional forms of the hard-dimer radial distribution function (Chiew, 1991; Yethiraj and Hall, 1992; Ghonasgi and Chapman, 1994; Sadus, 1999) were refit to a novel empirical functional form which results in an equation of state which is overall cubic in molar volume.

The resulting theoretically-based EOS was applied to predict several model shale gas properties including compressibility factor versus packing fraction isotherms, the vapor-liquid coexistence curves, the second-virial coefficient versus temperature and the self-diffusion coefficient versus pressure isotherms for the constituent pure fluids; and vapor-liquid equilibria and binary diffusion coefficients versus pressure for binary systems. The observed performance of the molecularly-based *cubic* EOS presented in this paper is generally better than the performance of conventional cubic equations of state and about equal to the performance of the more complex non-cubic SAFT equations of state for the same fluids but with much less computational effort. Ongoing work involves development of *cubic* EOS terms which account for multipolar and hydrogen bonding effect to include water and other species in the shale gas mixtures.

REFERENCES

Chapman, W. G., K. E. Gubbins, G. Jackson and M. Radosz, *New Reference Equation of State for Associating Liquids*, Ind. Eng. Chem. Res. **29**, 1709-1721 (1990).

Chen, S. S. and A. Kreglewski, *Applications of the Augment van der Waals Theory of Fluids: I. Pure Fluids*, Berichte der Bunsengesellschaft fur Physikalische Chemie, **81**, 1048-1052 (1977).

Chiew, Y. C., *Percus-Yevick Integral Equation Theory for Athermal Hard-Sphere Chains. II. Average Intermolecular Correlation Functions*. Molec. Phys. **73**, 359-373 (1991).

Ghonasgi, D. and W. G. Chapman, *A New Equation of State for Hard-Chain Molecules*, J. Chem. Phys. **100**, 6633-6639 (1994).

Huang, S. H. and M. Radosz, *Equation of State for Small, Large, Polydisperse and Associating Molecules*, Ind. Eng. Chem. Res., **29**, 2284-2294 (1990).

Peng, D.-Y. and D. B. Robinson, *A New Two-Constant Equation of State*, Ind. Eng. Chem., **15**, 59-64 (1976).

Redlich, O. and J. N. S. Kwong, *On the Thermodynamics of Solutions. V. An Equation of State. Fugacities of Gaseous Solutions*, Chem. Rev. **44**, 233-244 (1949).

Sadus, R. J., *A Simple Equation of State for Hard-Sphere Chains*, AIChE J., **45**, 2454-2457 (1999).

Soave, G. *Equilibrium Constants from a Modified Redlich-Kwong Equation of State*. Chem. Eng. Sci., **27**, 1197-1203 (1972).

van der Waals, J. D., *Over de Continuiteit van den Gas- en Vloeistoftoestand (On the Continuity of the Gas and Liquid State)*. Ph.D. thesis, Leiden, The Netherlands, (1873).

Wertheim, M. S., *Fluids with Highly Directional Attractive Forces I. Statistical Thermodynamics*, J. Stat. Phys., **35**,

19-34 (1984).

Yethiraj, A. and C. K. Hall, *Monte Carlo Simulations and Integral Equation Theory for Microscopic Correlations in Polymeric Fluids*, J. Chem. Phys., **96**, 797-809 (1992).

AUTHOR INDEX

AUTHOR INDEX